RECONSTRUCTING MATHEMATICS EDUCATION

Stories of Teachers
Meeting the Challenge of Reform

RECONSTRUCTING MATHEMATICS EDUCATION

Stories of Teachers
Meeting the Challenge of Reform

Deborah Schifter
Catherine Twomey Fosnot

Teachers College, Columbia University
New York and London

Published by Teachers College Press, 1234 Amsterdam Avenue, New York, NY 10027

The material in this book is based upon work supported by the National Science Foundation under Grants No. TEI-8552391, TPE-8850490, and TPE-9050350. Any opinions, findings, conclusions, or recommendations expressed in this book are those of the authors and do not necessarily reflect the views of the National Science Foundation.

Sections of Chapters 3 and 5 previously appeared in *Enquiring Teachers, Enquiring Learners: A Constructivist Approach for Teaching* by Catherine Twomey Fosnot, published by Teachers College Press.

Library of Congress Cataloging-in-Publication Data

Schifter, Deborah.
    Reconstructing mathematics education : stories of teachers meeting the challenge of reform / Deborah Schifter, Catherine Twomey Fosnot.
      p.   cm.
    Includes bibliographical references and index.
    ISBN 0–8077–3206–0 (alk. paper).—ISBN 0–8077–3205–2 (pbk. : alk. paper)
    1. Mathematics—Study and teaching.   I. Fosnot, Catherine Twomey.
II. Title.
QA11.S168   1992   92-30180
510′.71—dc20

ISBN 0–8077–3205–2 (pbk.)
ISBN 0–8077–3206–0 (cl.)

Printed on acid-free paper
Manufactured in the United States of America
99 98 97 96 95 94 93   7 6 5 4 3 2 1

To the teachers from whom we have learned so much

Contents

Foreword

Changing the way we teach and learn mathematics in school is an ambitious aim. Raised on mathematics as a body of rules, teaching as telling, and learning as memorizing, we cannot easily imagine classrooms that embody the sorts of principles and aims described in current reform documents. What counts as an authentic mathematical "task"? What does it mean to focus on "understanding"? What is there to discuss about mathematical questions and solutions? How can children learn what they are supposed to if teachers do not show them how? Envisioning classrooms in which students engage in mathematical problem solving, debate alternative solutions, develop connections and meanings, we do not understand enough about the challenge of trying to create such classrooms. This volume offers a rich anthology of the difficulties and joys, successes and struggles of a community of educators deeply engaged in efforts to realize those visions.

This book provides a valuable resource for anyone concerned with reforming mathematics education. For teachers, attracted to and challenged by the visions of reform, the stories of Lisa, Geri, Linda, Jill, and others will offer encouragement and inspiration. These are the stories of committed, talented—and real—teachers who are working their way through the disequilibrium, disorientation, and confusion of unsettling that which they had always done, and done well. These narratives offer opportunities to understand the dynamic and demands of change: the personal resources it takes, the time required, the complexity and confusion of inventing new practices. No one who reads this book can put it down thinking that "teacher training"—covered in the corners of the year, for brief mements of time, with little ongoing support and feedback—is what is needed. No one can think that teaching in these ways is a matter of acquiring a few new techniques, or that change can be accomplished efficiently, painlessly, or quickly. This work described in this book is hard. There are no straightforward paths. Invention, rather than implementation, is the task. In the words of one of the teachers, "It's scary!"

If this book has a moral, it is that support for teacher learning and change is essential if these kinds of changes in practice are to be realized. As Schifter and Fosnot note, teachers, themselves products of the very system they now aim to change, need opportunities to revisit and reconstruct their own understandings of mathematics. Teachers need chances to reconsider what it means to *understand* something in mathematics, and how such understanding can be fostered. Teachers need contexts in which thoughtful investigation of teaching is possible, where alternatives are imagined, discussed, tried, and assessed, and where such efforts are encouraged and supported. When new problems arise—such as exacerbating unequal patterns of access, participation, and benefit—teachers need support, not blame. They need encouragement not to give up, but to respond, when administrators or parents question the curriculum or what students are learning. These kinds of conditions for the development of their practice are far from commonplace for most practicing teachers. For those committed to creating meaningful change in how mathematics is taught and learned in classrooms, this book makes vivid what is possible when the demands of change are considered and when support of the hard work of change is ongoing.

Deborah Loewenberg Ball

Preface

Mathematics education in the United States is currently in a state of ferment: new standards and policies have been promulgated, new teaching materials are being prepared, and new assessment measures are being debated. But if all this activity is to result in a transformed mathematics instruction, it must be acknowledged—programmatically, in a serious commitment to teacher development—that, in the end, it is the classroom teacher who will interpret these standards and policies, who will decide how the new curricular materials are to be used.

In this book, we tell the stories of teachers who, guided by evolving constructivist understandings of mathematics learning, are meeting the challenge of reconstructing their teaching practice. We recount the struggles, questions, doubts, and successes they experience as they glimpse the possibility of a qualitatively different kind of school mathematics and then work to bring that vision to life in their day-to-day work with their students.

The reader is frequently brought into their classrooms to observe students engaged in mathematical activity: second graders hypothesizing about even and odd numbers; third graders demonstrating the commutative property of multiplication; sixth graders puzzling over the mysteries of fractions. In each situation we consider the teacher's intentions in designing the activity, the instructional decisions she makes as the children engage in it, and her reflections afterward.

We also tell the story of an in-service teacher education program, SummerMath for Teachers (see Appendix), committed to the principles of constructivism. We describe the staff's goals and assumptions, and how they meet the decisions with which they are faced as they work to help teachers implement the new mathematics pedagogy.

Although the stories all reflect the experiences of a single in-service program, each exemplifies an aspect of the change process that we believe to be generic to teachers' efforts to transform their mathematics instruction. While the program embraces teachers of grades kindergarten through 12, we limit the focus of the book to elementary teachers. In this way, we feel, the mathematical issues will be accessible to a wider audience. The fact that all teachers we write about are women is a reflection of the low percentage of male elementary teachers. (Less than 8% of the K-6 program participants are male.)

The data sources used in composing these stories of teacher development include teachers' own journals and other writings, audiotaped interviews, videotapes of classrooms, and notes kept during the year in which we provided follow-up support. The teachers agreed to be subjects of case studies, and each has read the chapter written about her. They were interested in how they were portrayed; some found it strange to read about themselves from someone else's perspective; a few were a little embarrassed, but all were willing to let their stories be told. All wanted their real names used.

The real names of SummerMath for Teachers staff are also used, but pseudonyms are employed for all students and for those teachers whose names do not appear in chapter titles or epigraphs.

Readers will find that some chapters require slower, closer reading than others. Following someone else's mathematical thinking, or pursuing one's own line of mathematical thought, requires a different kind of attention than reading about pedagogical issues or learning processes. For Chapter 3 especially, where readers are asked to explore an extended set of mathematical questions, it may be helpful to find a partner and work through some of the activities together. In any case, thoughtful examination of the mathematical issues will make clearer how the goals of the program animate the experiences of the teachers.

However, the chapters of this book do not have to be read in order. Readers who, at a given moment, are unable to commit the necessary time and thought to the mathematics involved might prefer to skip over those portions that require closer reading and return to them later.

The ideas expressed in this book have not been developed in isolation. The SummerMath for Teachers staff work collaboratively, which means that they meet frequently to plan, reflect, and analyze together. While they each bring their own perspectives, ideas are developed in common through discussion and debate. Of course, we, the authors, are ultimately responsible for the ideas expressed here, but our past and current staff—Virginia Bastable, Alison Birch, Ricky Carter, Jere Confrey, Ellen Davidson, Marsha Davis, Joan Ferrini-Mundy, Jim Hammerman, Fadia Harik, Paula Hooper, Jill Lester, Marty Simon, and Gini Stimpson—are present in these pages.

Special acknowledgment is due to Marty Simon, who was the program's director when each of us joined its staff. Under Marty's leadership, SummerMath for Teachers initiated its work with teachers in their classrooms. His approach to teaching and the analysis of practice has profoundly influenced our thinking. Marty also deserves special credit for having provided the inspiration for using the Xmania story to frame an exploration of the properties of numbers (see Chapter 3) and for developing much of the assessment material—concepts as well as language—discussed in Chapter 10.

We are greatly indebted to Alan Schiffmann for the many hours he spent with the manuscript of this book, challenging us to clarify our ideas and helping us to develop more fluent prose.

We would like to acknowledge the helpful comments of Deborah Loewenberg Ball, Ricky Carter, Ron Narode, Barbara Scott Nelson, Marty Simon, the Mount Holyoke College faculty writing group organized and led by Rebecca Faery, and the members of the spring 1991 class of the Mathematics Process Writing Project, who read parts or, in some cases, all of early drafts of the manuscript.

Finally, many thanks for their work with us to Melissa Mashburn, Brian Ellerbeck, and Peter Sieger, our editors at Teachers College Press.

1
Introduction

Why do we cover so much information in one year? . . . Why is it that kids come to the fifth grade and they still can't subtract? . . . Look at a third grade curriculum and there's fractions and there's decimals and there's telling time and every year we teach the same thing but we never get out of introducing it. . . . Are we here to introduce things or are we here for kids to really master them and understand them? And what does it take to do that?

Sherry Sajdak
Chestnut Hill Community School
Belchertown, MA

On a May morning in 1989, Deborah Schifter, one of the authors of this book, is visiting Ginny Brown's third-grade class. As she enters the classroom, she sees that just about everybody is talking. Scanning the room for an adult, she spots Ginny sitting with a group of children, listening, nodding her head, as one of them gestures excitedly. Ginny nods once more and says something to the group before she moves on.

It's clear that though there are many centers of activity in Ginny's classroom, it is far from chaotic. Students are organized in groups of four, their attention focused on jars of jelly beans. As Deborah approaches a group to see what they are doing, one of the children volunteers, "We have to find out how many jelly beans are in the jar. See, there are six layers, and 15 jelly beans in each one." Deborah sees that indeed Ginny has layered jelly beans in Skippy peanut butter jars, each layer separated from its neighbors by cardboard.

Deborah moves on to another group and asks, "What are you guys doing?"

"The answer's 90," Kevin says gruffly, looking at Deborah with a challenge in his eye.

"Kevin, she's not interested in the answer," Jessie interjects, "she wants to know how we're thinking about it!"

Kevin looks at Deborah again and then decides to explain his solution. "Andrew and I did it this way: 15 + 15 + 15 + 15 + 15 + 15. We add it all together and get 90. Jessie and Janice did 15 × 6. It's the same thing, 6 layers of 15."

1

2

Reconstructing Mathematics Education

"But then I wanted to do it another way," Jessie adds. "So I tried 6 × 15. I keep getting 80, but that can't be right." Deborah checks Jessie's worksheet and sees that she had written

$$
\begin{array}{r}
15 \\
\times\ 6 \\
\hline
90
\end{array}
$$

first, correctly applying the conventional multiplication algorithm. But then, when she tried to compute

$$
\begin{array}{r}
6 \\
\times\ 15 \\
\end{array}
$$

she had lost track of which digits multiply which and had ended up with 80. "That can't be right!" Jessie repeats with vexation. "I have to think about this some more."

When Ginny flicks the light switch, signaling "discussion time," the children take their papers and sit in a semicircle on the rug in front of the blackboard. "Who wants to lead?" Ginny asks the class. "Karen? Okay." Karen goes to the board while Ginny sits down on the floor with the children.

Several students offer to share their solutions. Some explain with diagrams, some use blocks, some calculate using the traditional algorithm, and some demonstrate unique ways of pulling the numbers apart and putting them together again. The children listen and compare.

Before they finish the morning's lesson, Jessie presents her problem. "I know I'm doing something wrong, but I can't figure out what it is. I want the class to help." When she explains her calculation—

$$
\begin{array}{r}
6 \\
\times\ 15 \\
\hline
80
\end{array}
$$

—her classmates puzzle over it; her mistake isn't obvious to them. Finally, Maria says, "Oh, I see it," and she points out Jessie's error to everyone else.

After class Ginny takes a moment to chat with Deborah. "It's a little embarrassing," she says, "but I've learned so much math since I started teaching this way. The children show me more than I could ever see alone."

Ginny Brown has been a teacher for over 20 years, but she has not always taught this way. "I was a strictly traditional teacher. I showed them how to multiply and then gave them drill and practice sheets. I didn't know there was anything else to it."

Two years prior to the lesson Deborah had just observed, Ginny entered a mathematics in-service program. "I discovered there's a lot more to mathematics, to learning mathematics, and to understanding mathematics, and I learned that I could teach the children to understand mathematics, too."

WHAT DOES "UNDERSTANDING MATHEMATICS" MEAN?

Ginny's new instructional practice is based on a fundamental rethinking of what "understanding mathematics" means. Because a near-universal consensus has lately developed that mathematics education in the United States is seriously inadequate to contemporary needs, an alternative pedagogy—a new paradigm of mathematics instruction—long in gestation, has begun to find the support necessary to contest the one in place. Yet what could be more obvious than the proposition that mathematics ought to be taught for understanding? Don't computational proficiency, as evidenced in pages of correctly solved drill and practice sheets, and grade-level-appropriate test scores establish whether and to what extent students understand the mathematics they are being taught? To explore these matters further, let us return to Ginny Brown's classroom.

Deborah is seated with Jessie, Janice, Kevin, Andrew, and their jar of 90 jelly beans. They have been asked to share the jelly beans among themselves. They know they won't be opening their jar until after lunch, so now they need to figure out how many beans each child will get without actually distributing them.

"Ninety divided by four," Jessie says as she writes "90 ÷ 4 = ____" on a sheet of paper. Meanwhile, Andrew goes to the storage shelf for a set of base-ten blocks. He shows Deborah the one-centimeter cubes and the rods that represent 10 cubes joined in a row. (See Figure 1.1.) "See, look," he says to Deborah. "We can show 90 with these." He counts out nine rods, "10, 20, 30, 40, 50, 60, 70, 80, 90."

"Okay," Jessie now takes over, "we need to share them. One for you, one for you, one for you, one for me," she says, distributing the nine rods among the four of them. "One for you, one for you, one for you, one for me. And this one we need to trade." The ease with which the group performs the operation suggests that they have done similar problems before. There are now two rods in front of each child. Janice counts out 10 cubes to trade for the one remaining rod and deals them out. She goes around twice and then the problem registers. "Uh-oh. Something's wrong. It didn't come out even."

"You must've done it wrong," Kevin says. "Here, let me do it." He gathers up all the blocks and repeats the process. The group watches his actions carefully and, when he gets the same result, they sit quietly for a

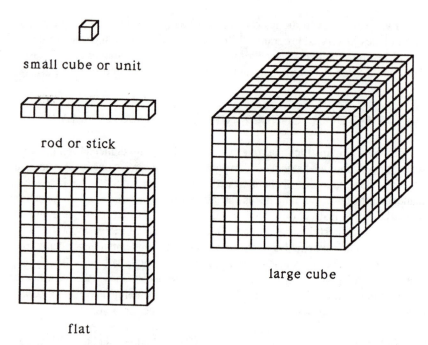

small cube or unit

rod or stick

flat

large cube

FIGURE 1.1. Base-ten blocks

moment. Then someone volunteers, "Maybe we traded wrong. Let's count out 90 cubes and do it without trading."

Until that morning, the sharing activities—division problems—that these children had encountered in class had all "come out even." Knowing that the activity required that each of them receive the same number of jelly beans, they assumed that something was wrong when they were confronted by a remainder. The problem Ginny's students had been asked to solve did not fit their current conception of what counts as division, and the unexpected result provoked feelings of confusion; in Piaget's words, they "experienced disequilibrium." At first they worked to resolve their disequilibrium by trying to find a mistake in their solution strategy.

Having counted out 90 cubes and dealt them, the children come up with the same puzzling result. "Two of us get 22 and two of us get 23. But that's not right! It has to come out even. It has to be fair!"

At this moment their teacher arrives to see what they are doing and the children explain their problem to her. "What do you think will happen when you open the jar?" she asks.

"The same thing! We'll each get 22 jelly beans and there will be 2 left over."

"And what are you going to do about that?"

Now once again imagining that they are dealing with jelly beans, rather than blocks, the children have some ideas. "We'll give one to you and one to her," Andrew suggests, pointing to Deborah. Simultaneously, Janice proposes cutting them in two so each child will get an additional half jelly bean.

They look at each other, and then look back at Ginny. "Which one is right?"

"Well, what do *you* think?" Ginny responds, knowing that once the children pose their own questions, they frequently find the answers for themselves.

"I think they're both right," Kevin decides. "It just depends on what you want to do."

Ginny looks over at the other group members. "What does Kevin mean?" she asks, not giving anything away.

"I get it," Janice says excitedly. "When you're dividing, it doesn't always have to come out even. But if it doesn't, you just have to decide what to do with what's left over."

Their confrontation with a novel problem—how to distribute 90 jelly beans among four children—a problem that did not initially fit their conception of division, had led these children to expand their notion of what constitutes division in order to "accommodate" their understanding to the new situation.

As Jessie looks down at the paper on which she has written "90 ÷ 4 = ____," she sees a new problem: "How do I write the answer?"

"How do *you* think?"

Jessie writes "22½" after the equals sign, and looks up. Ginny nods.

"What about the other way, when we give the extra jelly beans away?" Andrew asks.

"Then you would write it like this—'22 r 2,'" Ginny says, deciding that this is a matter of convention, not something they can reason out for themselves. "The r stands for remainder. You say, '22 remainder 2.' Can you explain why?"

"I know," Andrew says, pointing to what Ginny had written. "We each get 22 jelly beans and there are 2 left over—remainder 2."

When Janice and Jessie and Kevin and Andrew enlarged their conception of division, they were actively engaged in making meaning out of an initially puzzling experience. At the start of that lesson, each of these children was operating with a conception of division derived from situations, previously encountered, in which a given number of objects could be evenly distributed into a given number of groups, the distribution yielding whole numbers of objects without remainder. But the "jelly-bean problem" confronted them with a situation not assimilable to their current understanding, and the result was an experience of confusion, or disequilibrium. At first they tried to resolve that disequilibrium by looking for a mistake in their procedure, which would have left their conception of division intact. In the end, however, they resolved it instead by revising the conception (that division means partitioning a whole set into equal parts composed of whole numbers), in effect accepting a more inclusive understanding of what counts as division. Andrew suggested disposing of the extra jelly beans by giving them to non-group members, and Janice proposed cutting them up and distributing the pieces among themselves. Either way, once the children accepted that division could result in leftovers or fractional parts, they could assimilate the new possibility by adopting a more inclusive conception of the operation.

That Ginny's students were able to confront their own confusion about the problem and resolve it largely by themselves is a measure of the fluency with which they could move from written symbols to base-ten blocks to jelly beans and back. Working with mathematical concepts in the context of their own physical experience, the students could associate the words and symbols that designate division ("÷," "divided by," "remainder") with conceptual structures abstracted from the commonplace activity of sharing (distributing objects into equal-sized groups and deciding what to do with what is left over).

As these students' understanding of division grew to incorporate a wider range of conceptual structures, words and symbols that already denoted division for them were extended to these new structures, the resources of the notational system allowing them to generalize the results of their investigations. And their conception of division will continue to grow as they are confronted with new and different problems: by the end of the year, these third graders will encounter division as embracing at least two different physical operations—dealing a set of objects among a specified number of groups (as in the "jelly-bean problem") or partitioning a set of objects into groups of a specified size. And their conception of division will also include some sense of how that operation is related to addition, subtraction, and multiplication, how it might involve regrouping and distributivity, and how the calculations might be represented on paper. As these children interact in the rich mathematical environment of Ginny's classroom, the range and density of their "conceptual networks" will develop accordingly.

This picture implies that understanding is an ongoing process, deepening as conceptions expand and the number and solidity of connections among them develop. Strictly speaking, it does not make sense to say of a child (or an adult, for that matter), "She understands division," as though her understanding were complete. We listen to the child to find out which formulations she has made, which still remain unachieved, and which might next be explored to strengthen understanding further. With any mathematical concept, there are always new ways to look at it and more connections to be made.

Now contrast this sense of "understanding," as enacted in Ginny Brown's classroom, with that implied by the dominant paradigm. For example, in a conventionally conducted elementary-level arithmetic class, understanding has been evidenced, say, by correctly applying the addition algorithm for regrouping from the ones column to the tens column. Similarly, in the traditional algebra class understanding is demonstrated when students produce, through a succession of algebraic operations, the "right" string of expressions. And in the conventional geometry class, understanding a theorem means producing a valid two-column proof.

In Ginny Brown's classroom, the formal solution of a mathematical problem does not in and of itself demonstrate understanding. Thus, consider Susan, also a third grader but in a traditionally conducted class, who, when interviewed by a researcher, was confident in the division she was just now learning. Asked to solve a problem of her own devising, she proudly demonstrated how she could divide 24,682 by 5, following the long division algorithm carefully and correctly. The interviewer then asked her to divide 32 by 5, which, of course, she could do, but when asked to explain what "*r* 2" meant, she could only repeat "remainder two." When pressed, Susan looked bewildered. Changing tack, the interviewer asked her to show "32 divided by 5" with blocks. At first hesitant, she went on to count out 32 blocks and then began to distribute them into five groups. But as she neared the end of the process, she stopped, saying it couldn't be done. "It doesn't come out even." Asked whether the fact that three groups have 6 blocks but two have 7 might have anything to do with "*r* 2," Susan shrugged.

Susan's conception of division involves following a particular procedure on a set of written symbols and also includes distributing blocks into equal-sized groups. However, to Susan these two representations of division appear to be unrelated. On the one hand, "32 ÷ 5," performed on paper, yields "6 *r* 2." On the other, "32 ÷ 5," performed with blocks, has no meaning. Lacking any sense of their connection, Susan cannot use one representation to inform the other, evidence that her understanding of division is thin.

By contrast, Ginny Brown's third graders have constructed a rich set of conceptual linkages, one that, for example, provides an interpretation for the expression "*r* 2" that is rooted in a process they all understand well. Were they challenged to tell a story about sharing in which *r* would be *3* or *4*, they could do so.

This is not to say that all mathematical activity must retain an immediate relationship to the physical world. After enough experience distributing objects, the children in Ginny's class will be able to imagine sharing without having to enact it: reflection on their actions will allow them to generalize from specific cases to form more abstract notions of division (though reference to the "jelly-bean problem" can be called upon as needed). In just a few years, their under-standing of division of whole numbers will form the basis for constructing an understanding of division of fractions, and, later still, for algebraic rational expressions, hyperbolic functions, or the limit of $\frac{1}{x}$ as x increases without bound. If more abstract mathematical concepts are embedded in a network of solid constructions, the resulting richness of meaning will allow these concepts to be used flexibly and creatively to solve novel problems.

It is recognized as one of Piaget's central insights that if human understand-ing develops by devising categories and building conceptual structures for the interpretation of experience, this development is propelled by repeated phases of conceptual reorganization as established structures are revealed to rest on overgeneralized patterns or inappropriate analogies. Thus, the learning process proceeds through a succession of once serviceable but now no longer adequate notions, no matter what the context or who the teacher. For example, teachers recognize the common misconceptions that a multiplicative product is always larger than its factors and a quotient smaller than the dividend; that the sum of two or more fractions is obtained by adding the numerators and adding the de-nominators; that of two figures, the one with the larger area also has the larger perimeter. And in the present instance, both Susan and Ginny Brown's students assume that division always implies that the distribution of objects "comes out even." Of course, such beliefs, although incorrect, have in them an element of logic. For example, that a multiplicative product is larger than its factors is true of whole numbers (excepting 0 and 1), the context in which students initially make this conjecture. But when they move to the study of fractions and confront a class of exceptions to that rule (for example, $\frac{1}{2} \times \frac{1}{2} = \frac{1}{4}$ and $\frac{1}{4}$ is less than $\frac{1}{2}$), they find themselves in a state of disequilibrium. In general, such apparent contradictions are resolved by developing finer distinctions and more encom-passing understandings—in this case that the product is larger only when the factors are greater than one.

In sum, the construction and subsequent elaboration of new understandings is stimulated when established structures of interpretation do not allow for sat-isfactory negotiation of an encountered situation. The disequilibrium attendant upon failure to understand or to anticipate leads to mental activity and the mod-ification of previously held ideas to account for the new experience (Simon & Schifter, 1991).

Because this perspective on what is meant by "understanding mathematics" derives from the view that learning is primarily a process of concept construction and active interpretation—as opposed to the absorption and accumulation of received items of information—it is referred to as "constructivism." While this core principle of constructivism is not new, it has in recent decades been greatly elaborated, both theoretically and empirically (Kamii, 1985; Labinowicz, 1980; Piaget, 1972, 1977; von Glasersfeld, 1983, 1990). In particular, the adoption of this perspective by cognitive researchers has resulted in considerable insight into how mathematics is learned (Davis, 1984; Ginsburg, 1977, 1986; Hiebert, 1986; Resnick, 1987; Silver, 1985; Steffe, Cobb, & von Glasersfeld, 1988).

A NEW PARADIGM OF MATHEMATICS INSTRUCTION

If the creation of the conceptual networks that constitute each individual's map of reality—including her mathematical understanding—is the product of constructive and interpretive activity, then it follows that no matter how lucidly and patiently teachers explain *to* their students, they cannot understand *for* their students. Though this is a truism, it nonetheless has profound implications for pedagogical theory and practice: Once one accepts that the learner must herself actively explore mathematical concepts in order to build the necessary structures of understanding, it then follows that teaching mathematics must be reconceived as the provision of meaningful problems designed to encourage and facilitate the constructive process. In effect, the mathematics classroom becomes a problem-solving environment in which developing an approach to thinking about mathematical issues, including the ability to pose questions for oneself, and building the confidence necessary to approach new problems are valued more highly than memorizing algorithms and using them to get right answers (Ball, in press-a, in press-b; Cobb, 1988; Confrey, 1985, 1990; Davis, Maher, & Noddings, 1990; Duckworth, 1987; Fosnot, 1989; Lampert, 1988a; Simon, 1986).

When teachers are no longer sources of mathematical truth, but creators of problem-solving situations designed to elicit the discovery of new conceptual connections and new understandings, then it is especially important that they select tasks whose completion requires a cognitive reorganization. Thus will their students be called upon to use extant knowledge in new ways and to alter or expand their current stock of understandings.

To give these abstractions more substance, let us consider again Ginny Brown's class. Recall that the children were put into groups of four, each provided with its jar of jelly beans, and were then challenged to determine how many jelly beans each child would get upon distribution of a jar's contents. Ginny had designed an activity in which her students could work from what they were confident they understood—that division involves distributing a set of objects into a given number of groups—but she had introduced a new issue—that the operation would not result in whole numbers without remainder—in order to stimulate a transformation of their initial conception.

The structure of the class—students organized as groups of four—facilitated learning in a number of ways. For one, by working in a setting in which they verbalized their own ideas, students could clarify their understanding and identify confusion. The value of verbalization is often recognized by teachers who say, "I really began to understand the material only when I had to explain it myself." In Ginny's classes, that opportunity is not reserved to the teacher alone, but is shared by everyone.

The group setting also guarantees that different perspectives will be presented. As students consider one another's ideas, they are challenged to assess the validity and consistency of all those ideas. As one teacher suggested, reflecting on her own experience learning mathematics in small groups,

> [Working in a group] positions an idea in several different contexts simultaneously, as each person discusses the idea as they see it. Our pool of ideas becomes more expansive than it could ever be with each of us working alone. Working in such groups creates the feeling that together we each become better than we are alone, inspiring one another and challenging one another to defend our solutions or expand them to encompass others' ideas as well as our own.

And, in the words of a sixth grader, "I have an idea and my partners have their ideas. We listen to each other and put them together so then we all have one big idea."

When a class is run the way Ginny Brown's was, it is not at all a matter of "anything goes as long as the kids are having fun." While her students were given considerable leeway to explore those aspects of the problem that interested them, their teacher had carefully structured the context and chosen her questions in order to highlight the mathematics she felt they needed to learn. As Ginny's students explored various properties of the division operation, she attentively monitored their work, using her comments and questions to guide them.

On the other hand, Ginny's interventions were not designed simply to lead her students to the "right" answer. Rather, they were meant to encourage deeper understanding by drawing out a variety of solution strategies. Ginny exemplifies the teacher who, by the skillful use of nondirective but probing questions, facilitates reflection on, and verbalization of, her students' underlying thought processes. Such questioning invites students to identify the sometimes subtle choices they are making and articulate their reasons for making them.

As Ginny moved from group to group, she listened for the reasoning her students were applying, and, based on what she heard, she would pose additional questions in order to throw that reasoning into relief. She was especially alert to any misconceptions her students might reveal. Because she appreciated the developmental possibilities those misconceptions represented, she sought to evoke experiences that would lead her students to confront them, and, by doing so, to help them to new, more powerful and more discriminating cognitive structures. Between her monitoring of small-group explorations and the whole-group discussions that usually ended her lessons, she had sufficient information on which to base her decisions about subsequent directions.

The mathematics classroom focused on problem solving and conceptual understanding, rather than on computational drill, promotes students' confidence in their own mathematical abilities. As they discuss, debate, and challenge one another's ideas, a shift occurs in the locus of authority over both the material to be learned and the process of learning it. When students' ideas are taken seriously by teachers and classmates, they learn that those ideas are valuable. A classroom no longer dominated by the fetish of the "one right way"—the teacher's way, the textbook's way—to solve a problem has become a community of inquiry and discourse whose members explore the mathematics together. As a matter of principle as well as pedagogical strategy, when students pursue alternative paths to a solution, those paths must, if logically sound, be validated precisely because *they* pursue them.

Doing mathematics can be pleasurable, even exhilarating, but it does not always feel good. Being challenged, encountering novelty, confronting one's misconceptions—in short, building new and stronger understandings—typically involves bewilderment and frustration. However, if mathematics students can learn to recognize that the discomfort they experience is part of the process, they can also learn to tolerate it. By encouraging students to monitor their own learning, teachers can help them achieve greater control over that process. At the end of the school year, one third grader wrote to her mathematics teacher, "Whenever my partner and I were confused, I'd raise my hand and you'd come over and ask us questions and make us explain. That's how we learned and understood."

While the new paradigm frames a consensus on what, in general terms, "promoting understanding" means, the specifics of classroom technique and strategy depend on the particularities of the teacher, the student, their relationships, and the material to be taught. It is worth underscoring that no technique will in and of itself necessarily lead to successful learning. If Ginny's students developed rich mathematical conceptions working in groups of four, it was not the fact of group work alone that was responsible. Her students had had prior experience in group learning. They had entered her classroom with a sense of mathematics as a realm to be explored and, if explored, comprehended. They had also learned how to ask their own questions, and they did not fear that their thoughts and ideas would be ridiculed. In the absence of factors such as these, group work might have proven just another failed gimmick.

Implicated in these discussions of what is meant by "understanding mathematics" and "teaching mathematics" is a view of the nature of "doing mathematics" that differs sharply from that associated with the reigning instructional paradigm. In contrast to the static picture of the discipline—"all the math there is is already out there"—conveyed by the math-fact, drill-and-practice approach, the new view recognizes the reality that mathematics is a human pursuit with a long history: schools of thought compete, fashions change, and some questions may never be settled. If anything, the tremendous proliferation of new areas of mathematical research in our century has undermined rather than confirmed received ideas about the nature of mathematical truth. In earlier centuries it was the apparent certainty of mathematics that raised it above the natural sciences. Today, a keener appreciation of the interplay of imagination and logical necessity, a greater awareness of the roles of convention, philosophical commitment, and technological interest in shaping the development of the discipline, favors an emphasis on similarities between mathematics and the natural sciences. It is in this spirit that a pedagogy informed by the constructivist perspective aspires to convey at all levels of instruction an experience of mathematics approaching that of the mathematician. Posing questions, making and proving conjectures, exploring puzzles, solving problems, debating ideas, contemplating the beauty of mathematical structures—these constitute the "doing of mathematics" (Davis & Hersh, 1980; Ernest, 1991; Lakatos, 1976; Lampert, 1988a; Scott-Hodgetts & Lerman, 1990; Tymoczko, 1986).

RECONSTRUCTING MATHEMATICS EDUCATION: THE BARRIERS TO REFORM

The numerous recent calls for mathematics education reform in the United States tend to converge around a vision of the mathematics classroom evoked by the new paradigm—a vision much like that exemplified in Ginny Brown's third-grade classroom in western Massachusetts (California State Department of Education, 1985; Mathematical Association of America, 1991; National Council of Teachers of Mathematics, 1989, 1991; National Research Council, 1989, 1990). As members of the mathematics education community—teachers, administrators, authors of textbooks and designers of software, teacher educators and researchers—consider the implications of all the various frameworks, standards, and guidelines, their hopefulness must be tempered by recognition of the scope and complexity of the transformation they are being called on to help bring about. Mary Kennedy, director of the National Center for Research in Teacher Education at Michigan State University, has characterized the history of American education as "a history of reform efforts, most of which have left teaching unchanged" (1991, p. 3; see also Cohen, 1988; Cuban, 1984; Sarason, 1971, 1990).

Contemplated in the light of that history, the gap between extant instructional practice and the new paradigm being urged may appear too vast to bridge.

In fact, it is not difficult to find reasons why the current effort should meet with the same unhappy fate as its numerous predecessors.

1. *If teachers are necessarily the agents of this transformation, they are also the products of the system that they are being asked to change.* What they have learned about teaching they derive, first of all, from at least 16 years of experience as students—a longer internship than that undergone in any other profession. Thus, even before they enter certification programs, prospective teachers are already familiar with a wealth of instructional approaches and particular techniques, and have developed fairly well-defined, if usually unconsciously held, notions of the nature of the disciplines they are to teach. A pedagogy that centers on conjecture, conceptual exploration, and discursive interchange must combat a well-entrenched pedagogical practice that emphasizes memorization and computational routine conveyed through lecture, demonstration, or textbook (Cohen, 1988; Fosnot, 1989; Kennedy, 1991).

2. *What is being asked of teachers is not only very different from what they have done in the past, it is also much harder.* With respect to content knowledge, teachers must have an understanding of the mathematical concepts they are charged with teaching, including a sense of the connections that link these concepts to one another and to relevant physical contexts. In addition, teachers must know something about how students construct mathematical concepts so that they can follow their students' reasoning and, by skillful questioning, help them to extend their understanding. And finally, teachers must be able, like Ginny Brown, to select tasks that are grounded in what students already know, but that give them access to new concepts (Cohen et al., 1990; Shulman, 1986a, 1986b).

With respect to classroom process, teachers will have to manage a much more complex set of interactions. In traditional classrooms, they are broadcasters of information (almost all communication is initiated from the front of the classroom) and their students are required to demonstrate that they have received the message. Now, teachers must learn to disperse their authority to initiate and carry on discussion. While their students are working in groups, they must determine, as they visit each group, what the issues are and what type of intervention, if any, is appropriate. And when teachers facilitate whole-group discussion, they must constantly make decisions about how to focus the discussion—whose, or which, ideas to pursue—so as to maintain a balance between the interests of their most vocal students and their own instructional agenda (Ball, in press-a, in press-b).

3. *Most teachers rely heavily on the textbook for their mathematics lessons.* But traditional texts embody a transfer-of-information, drill-and-practice approach to instruction and are of limited use, except perhaps for reference purposes, in the kind of instruction we are proposing. Since there is very little alternative material available, teachers are left with the enormous burden of

redesigning their own mathematics curricula (Ball & Feiman-Nemser, 1988; Schram, Feiman-Nemser, & Ball, 1990).

4. *Most schools continue to place considerable emphasis on standardized tests designed to assess speed and accuracy in computation.* Because teachers' work is often evaluated on the basis of their students' test scores, even those teachers who are skeptical of the value of standardized testing are justifiably reluctant to wander far from computation-centered lessons when computational speed and accuracy are, after all, what these tests measure (Carl, 1991; Mokros, 1991). Of course, where "mathematical understanding" *means* speed and accuracy—and many teachers accept the equation—the inference that these tests truly reflect that understanding is unexceptionable.

5. *Even when teachers develop new conceptions of what it means to learn mathematics, they are, in general, working within a culture in which good teaching is assumed to mean ensuring that students get right answers.* This entails, first of all, that when students enter their classrooms in September, they expect their teachers to behave in accustomed ways. Should those expectations be thwarted, students will generally experience confusion, frustration, and sometimes even anger. At times, they actively resist new instructional approaches. (Such reactions tend to be more prevalent in older grades.)

If the teacher perseveres and is able to establish a new culture in her mathematics classroom, frequently parents become concerned that their children are not bringing home pages of computation each evening. Some are suspicious of manipulatives, diagrams, or other alternative representations (which they often refer to as "crutches" or believe are suitable only for very young children) and are concerned that their children are at risk if the teacher is experimenting with new methods. Parents who themselves excelled in school are particularly anxious to see their children race through the upper-grade curriculum and are not aware that there are other levels of understanding or other aspects of mathematics to explore. "What was good enough for me is good enough for my child" is what they often say.

6. *It is also often the case that supervisors and administrators who claim to embrace this new approach interpret it in terms of past reform movements whose premises are in conflict with this one.* For example, they may encourage their teachers to enact what they learned in an in-service program based on the new teaching paradigm (which administrators frequently reduce to the latest in fashionable strategies—using manipulatives, say, or group problem solving), but when they visit classrooms for purposes of evaluation, they look for goal-directedness, coverage, and closure—criteria generally inappropriate to an instructional practice centered on problem-generated mathematical exploration (Lampert, 1988b).

IMPLICIT IN THIS ENUMERATION of some of the more daunting barriers to the reform of mathematics education is the conclusion that much more is involved than simply tinkering with a basically sound system—adjusting here, adding on there. The system *is* the problem.

CONSTRUCTING THE NEW PARADIGM

In 1985, California's State Department of Education issued a policy declaration, *Mathematics Curriculum Framework for California Public Schools,* which foreshadowed more recent nationwide calls for reform in its privileging of mathematical understanding over computational proficiency. California's educational policymakers sought to use the new *Framework* to press publishers for more innovative mathematics texts—the major vehicle envisaged for implementing the new curricular guidelines. But when, in 1989, a group of researchers at Michigan State University (MSU) sought to assess the impact of the new policy on California's elementary schools, they found it had been at best superficial (Cohen et al., 1990). Because underlying beliefs about learning and about the nature of mathematics were not being confronted, the MSU study argued, it seemed that even as teachers experimented with new mathematics concepts and instructional strategies, most did so within the ambit of the old paradigm.

> So far teachers have been asked to make great changes, but they have not been offered many of the resources that might support such change. Few teachers have had opportunities to see examples of the sort of teaching that the state thinks it wants. Few have been offered opportunities to learn a new mathematics. Few have been given opportunities to cultivate a new sort of teaching practice, and even fewer have been offered assistance in the endeavor. (p. 165)

To the MSU team, the California experiment suggested that in the absence of a serious teacher development initiative, bold policy directives and expertly produced curricular materials would not, by themselves, realize the vision proposed in the *Framework*. "In a word," their study concluded,

> teachers have not yet been engaged in the sort of conversation—with themselves, with other teachers, with university mathematicians and many others—that would support their efforts to learn a new mathematics, and a new mathematical pedagogy. (p. 165)

Though still few in number, there are teacher education programs in the United States, both pre- (for example, see Fosnot, 1991; McDiarmid, 1990; Simon, 1991) and in-service (for example, see Carpenter & Fennema, in press; Hart, 1991; Maher & Alston, 1990; Mumme & Weissglass, in press; Russell & Corwin, 1991; Wood, Cobb, & Yackel, 1991), already engaging teachers in just that sort of conversation. Moreover, the collective experience of these programs argues an even stronger conclusion than that drawn in the MSU study: given the vision of mathematics instruction animating the reform effort, teacher development—teachers constructing for themselves the new mathematics pedagogy—is at the heart of that vision.

SummerMath for Teachers, in existence since 1983 and based at Mount Holyoke College, was among the first in-service programs to introduce teachers to a constructivist perspective on mathematics education. As it has developed over the past decade, the program's work has come to be guided by four basic principles:

1. *The approach to learning and teaching mathematics presented in this introduction translates to learning and teaching in general. Thus, teacher education should be based on the same pedagogical principles as mathematics instruction.* There are no recipes for ensuring that teachers develop a classroom practice like Ginny Brown's. Simply providing them with a repertoire of teaching strategies and techniques would be equivalent to teaching Susan the long-division algorithm without helping her to understand why it works so that she knows when to use it and for what purpose. In the end, the classroom structures as well as the materials Ginny chooses and the questions she asks are all subordinated to the mathematics she wants her students to learn and her understanding of how they will best learn it. The subtleties of timing and tone and the appropriateness of her spontaneous responses stem from a coherent conception of what should be happening in her classroom.

A course of lectures on "constructivist perspectives on mathematics instruction" is certain to be equally unproductive. Were such concepts presented in lecture format, most teachers would keep the ideas at arm's length, dismissing them as wrong or inapplicable, or interpreting them in a way that minimizes their impact. Even for those teachers who might be eager to engage them seriously, it would be next to impossible to translate such abstract notions into the nitty-gritty of day-to-day instruction.

Rather, teachers must be given experiences as students in classrooms that embody the new teaching paradigm. Such classes must provide learning experiences powerful enough to challenge 16-plus years of traditional education. Teachers must be able to recognize that this is the kind of learning that they would choose to foster in their own classrooms, and they must be given opportunities to critically analyze the process of learning, the nature of mathematics, and the kinds of classroom structure that will promote that goal.

2. *If teachers are expected to teach mathematics for understanding—in the sense explicated here—they must themselves become mathematics learners.* In the usual teacher education course, hands-on experience means trying out lessons designed for schoolchildren, with the result that even teachers with an inadequate mathematics background experience little or no conceptual develop-

ment or expansion of problem-solving skills. But when teachers are challenged at their own level of mathematics competence, are confronted with mathematical concepts and problems they have not encountered before, they both increase their mathematical knowledge and, more importantly, experience a depth of learning that is for many of them unprecedented. Such activities allow teachers, often for the first time, to encounter mathematics as an activity of construction, of exploration, of debate, of a complex interplay of convention and necessity, rather than as a finished body of results to be accepted, accumulated, and reproduced. Experiences like these are by far the most persuasive impetus for a changed mathematics pedagogy and serve as vital touchstones of what is possible when the uncertainties and frustrations of change threaten to overwhelm teachers' best intentions.

Teachers are often surprised to discover that the topics of the elementary school mathematics curriculum provide rich opportunities for challenging mathematical exploration. By digging more deeply into what, on the surface, seems like familiar territory, they develop a broader sense of the conceptual issues their students confront.

As in-service programs begin to provide such opportunities, they will empower teachers with the understanding and the authority to make instructional decisions without relying exclusively, or even primarily, on the textbook. As better instructional materials become available (for example, Bennett, Maier, & Nelson, 1988; Burns, 1987, 1991; Corwin & Russell, 1990; Russell, in preparation; Shroyer & Fitzgerald, 1986), teachers should be in a position to determine for themselves which materials to use, how to use them, and when.

3. *Regular classroom consultation provides support for continued reflection as changes are introduced into the classroom, sustaining teachers' learning in the context that matters most.* Translating the insights gained in an in-service course into new classroom methods is always a difficult, and often a frustrating, process. Pressure from administrators and parents to cover an existing curriculum, lack of institutional support, resistance from students, and other demands on teachers' time may all reduce the actual effect that in-service programs have on instruction. Implementation efforts may be put off indefinitely. Initial efforts that do not meet with instant success (the norm rather than the exception) are often abandoned. A more profound and longer-lasting impact can usually be realized only when programs integrate clinical supervision of classroom practice with courses and institutes (Joyce & Showers, 1988; Simon, 1989).

Particularly important, given the type of instructional transformation proposed, are the balance and perspective that an experienced observer can offer to a teacher who is struggling to implement new classroom goals. For example, a teacher who has previously judged a lesson successful if in his quiet and orderly classroom students are able to answer his questions correctly is likely to be discomfited when they disperse into groups, raising the noise level as they wrestle with their assignment. A classroom consultant can help the teacher to see the learning that is taking place.

Of course, the classroom is an incomparably rich environment for continuing to learn about learning. Each day students' thoughts, questions, and behaviors offer new information to be incorporated into an evolving instructional practice. One major function of the classroom consultant is to alert teachers to particular events and behaviors whose significance they might not appreciate and to stimulate reflection on these, in order to help teachers become more self-conscious about their own decision-making processes. Follow-up support can also provide the reassurance necessary if such opportunities for deeper reflection and guidance in the integration of theory and practice are to be exploited.

4. *Collaboration among teachers is essential to the process of reform.* The workaday experience of most teachers is one of isolation: classroom doors are closed and, while the few moments of daily collegial interaction allow for venting frustration over a particular problem student or a new piece of bureaucratic folly, they don't permit serious discussion of instructional issues.

In making so little provision for teachers to reflect together on their instructional practice, school structures support a culture of isolation consistent with the reigning paradigm. For, when instructional initiative is wired into district-wide curricular guidelines and text- and workbooks, the important decisions are largely out of teachers' hands and so they don't need to trade ideas on either their teaching or mathematics. But the kind of transformation proposed here, in investing greater instructional responsibility in the teacher, concomitantly entails greater need for collegial cooperation. In the long run, only teachers taking similar risks and experimenting with similar instructional approaches are in a position to support one another. They can, and must, share their reflections on classroom process, help one another plan appropriate lessons, and explore together the mathematics they teach (Weissglass, 1991).

No less important than these instructional issues are the institutional realities that must be brought into line with the new mathematics pedagogy—for example, curriculum development, the selection of texts and other teaching materials, standardized testing, parent and community outreach, and, not least, new measures of teacher effectiveness.

In-service programs will have to recognize that encouraging the development of collaboration among their participants is as integral to their efforts as introducing new models of instruction. As such programs mature and past participants emerge as educational leaders, the latter become the most effective promoters of change, as much in school- and district-wide policy as in their colleagues' classrooms.

THE 10-YEAR HISTORY of SummerMath for Teachers (see Appendix) demonstrates that though the barriers to reform are formidable, they are not insuperable. Admittedly, not all who participate in the program make significant changes in the way they teach mathematics, but many do transform their practice in fundamental—and, for them, exciting and rewarding—ways. In telling the stories of nine teachers who have engaged in constructing a new pedagogical practice, this book describes some of the kinds of changes they have gone through.

Initial chapters illustrate how the process begins with in-service course-work. Chapter 2 follows three teachers, Ana Malavé, Betsy Howlett, and Pat Collins, through a summer institute designed to foster the development of new visions of classroom practice. Chapter 3 describes a lesson in which 36 teachers are invited to explore the properties of different systems of number notation, leading them to discover for themselves the power of the idea of place value and the role of choice and invention in mathematics. The story of Linda Sarage, told in Chapter 4, is also an account of a full-semester mathematics course in which teachers revisit the mathematics topics they teach, exploring the conceptual connections—the "big ideas"—that underlie and unite those topics.

Once back in their own classrooms, teachers find that big changes generally do not come either quickly or smoothly, but occur in stages, by fits and starts, over the course of months and even years. During a summer institute or semester-long course, teachers develop new ideas about learning and teaching and plan ways of implementing new techniques. Some return to their students with a neat list of strategies based on a rather shallow conception of how learning takes place. Others take with them both a much richer sense of learning and a great deal of confusion about how it applies to classroom process.

Either way, whether confident or tentative, teachers' initial attempts at applying what they have learned are typically disjointed, and management problems are common. But if teachers persevere, classroom management becomes routine, and they are free to direct their attention to student learning. Follow-up consultants provide them with opportunities to reflect upon classroom process, helping them develop further their ideas about learning and teaching. Jill Lester's story, told in Chapter 5, exemplifies this process.

Nor is change easy. Teachers are frequently frustrated, at times angered, by program experiences that create disequilibrium, thereby challenging their professional identities. ("I always thought I was a good teacher because *I* explained things so clearly. Now I see how I've cheated my students out of the opportunity to explain.") Lessons drawn from their pre-service years, from their own classrooms, from administrators' directives, and from textbook curricula are usually all at odds with what they experience at SummerMath for Teachers and some teachers have considerable difficulty coming to terms with the clash of perspectives. Chapter 6 describes the persistence with which Sherry Sajdak negotiated this painful process.

Yet despite the inevitable emotional turmoil, the rewards appear to be worth the effort. For teachers who have a history of failure and mathematics avoidance, experiencing their own effectiveness as mathematical thinkers in classes taught by program staff can prove especially exhilarating. Chapter 7 tells how Lisa Yaffee's development as a mathematics learner allowed her to transform her mathematics instruction.

While the focus of our work with teachers is obviously on developmental change in instructional technique and conceptions of classroom process, we have repeatedly seen how participation in the program affects teachers' lives well beyond the confines of the classroom. For many, their work has always involved following someone else's directions: in college, they heeded their instructors; as student teachers, they listened to their supervisors; and now, as practicing teachers, they have school administrators, in-service program instructors, and textbooks telling them how and what to teach. At SummerMatch for Teachers, when they are challenged to develop and articulate their own ideas, they learn not only about understanding and teaching mathematics but also about their own powers of thought. Thus, Chapter 8 describes how Ginny Brown acquired a new sense of self, conquering insecurities and daring to take risks she would not have contemplated before. As a consequence, she entered a graduate program in education where she learned that the basic principles governing learning and teaching mathematics also applied to language arts and science.

Not only do these principles apply across disciplines; they also apply to people of all ages. As past participants in SummerMath for Teachers become educational leaders, carrying on in-service work in their schools, they find that the premises on which their new approach to mathematics instruction is based are valid for their work with their colleagues. Chapter 9 follows Geri Smith as she developed a new sense of what it means, first, to be a teacher, and then, to be a leader in her school and district.

In this book we have chosen to write about teachers who, in engaging seriously the habits and assumptions that have defined their practice as educators, in some cases for as much as 20 or 30 years, are reconstructing mathematics education. With these stories of success, we mean to show that such profound, albeit difficult, changes are not only possible but deeply satisfying, both professionally and personally. If these teachers are perhaps more articulate than some, or their experiences more extreme, their stories nonetheless represent the sorts of changes required if the challenge of mathematics education reform is to be met. It is our hope that in telling these stories, we can convey enough about the process of change to help support others through it.

Part I

TEACHERS AS STUDENTS

2
Cultivating New Visions:
Ana Malavé, Betsy Howlett,
and Pat Collins

The likelihood that professional study will affect what powerful early experiences have inscribed on the mind and emotions will depend on its power to cultivate images of the possible and desirable and to forge commitments to make those images a reality.

Sharon Feiman-Nemser, "Learning to Teach"

Teachers teach according to some conception—most often implicit in what they do, less often explicitly held—of how learning takes place and of the nature of the content they teach. While this may seem obvious, it does suggest a starting point for teacher development: challenge those conceptions, open them up to a process of reflection. With the occasion thus afforded teachers for gaining deeper insight into instructional theory and practice, new visions "of the possible and desirable" can begin to take shape.

Yet, such insight, if it is to result in significant and enduring change in the way teachers teach, cannot be induced by a course of lectures, a handful of workshops, or even books like this one, no matter how informative or persuasive. Instead, teacher development programs will have to dig deeper, furnishing their participants opportunities to construct for themselves more powerful, alternative understandings of learning, teaching, and disciplinary substance. To this end, in-service activities should be designed to provide experiences that cannot be smoothly assimilated with dominant instructional paradigms. And teachers must be invited to confront and work through such experiences if they wish for themselves a more coherent and personally compelling practice. At the same time, activities capable of such impact will also have to be rich enough and flexible enough to engage each of the teachers in accordance with their varying backgrounds and histories.

In this chapter three teachers are followed through the two intensive weeks of a SummerMath for Teachers institute: Ana Malavé, who conducts her instruction for her urban first graders in their and her native language, Spanish; Betsy Howlett, who teaches sixth graders in a rural area; and Pat Collins, also a sixth-grade teacher, but in the laboratory school of a private liberal arts college. The institute, attended by 36 elementary teachers, is staffed by Deborah Schifter, Jim Hammerman, and Ellen Davidson.

LEARNING ABOUT LEARNING THROUGH LEARNING MATHEMATICS

"Here's what I think about the problem," says Norma, a primary-grade ESL teacher from Holyoke, Massachusetts, a bit self-consciously. She and her classmates are working on the following problem: *It takes Sam 2 hours to dig a ditch and it takes Terry 4 hours. If the two of them work together, and efficiently, how long will it take them to dig the ditch?*

"I think that if Sam and Terry are anything like my children, Sam would dig his half in 1 hour and then sit down until Terry finishes his part. It would take Terry 2 hours to finish, so the answer is 2 hours."

While the class laughs and Norma expects her answer to be dismissed as a joke, Jim, who is leading the discussion, responds quite seriously. "Norma, you've made certain assumptions about the working relationship between Sam and Terry. Is your answer valid, based on those assumptions?"

"I think so," she answers hesitantly.

"But wait a minute," declares Betsy. "It says here that they worked efficiently. I take that to mean that they both worked until it was done. So the answer has to be less than 2 hours."

"Hmmm," responds Jim. "And so how did you solve the problem?"

Betsy stands to show the class her solution, using the Unifix cubes. (See Betsy's solution in Figure 2.1.) "Let's say these 16 cubes stand for the whole ditch. In 1 hour, Sam digs 8 cubes' worth and Terry digs 4 cubes' worth." She holds up the 16 cubes, breaks off 8 and then 4 more, putting them in her right hand, and displaying the remaining 4 in her left. "Now we have these 4 cubes left. Since Sam digs 8 cubes' worth in 1 hour, then we know he digs 2 cubes' worth in 15 minutes. Terry, who works half as fast, digs 1 cube's worth in that time. So 15 cubes' worth is dug in 1 hour and 15 minutes." She pauses and looks around the room. When she sees a few people nodding their heads, she continues. "Now, what do we do with that last cube? Well, one of these cubes took

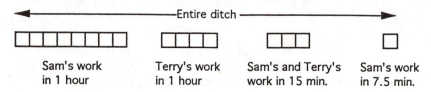

FIGURE 2.1. Betsy's solution to the "ditch problem"

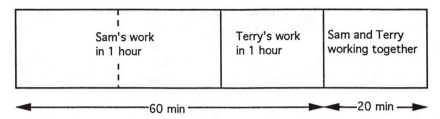

FIGURE 2.2. Cora's solution to the "ditch problem"

Sam 7½ minutes to dig, so we have to add on another 7.5. That's 1 hour and 22.5 minutes altogether." She sits down and waits for a response.

"That can't be right," objects Cora, a fourth-grade teacher from a private school in Texas.

"Why not?"

"Well, this is what I did." Cora walks up to the chalkboard at the front of the room and draws a rectangle. (See Figure 2.2.) "This rectangle stands for the whole ditch. In 1 hour Sam could dig ½ of the ditch." She divides the rectangle in half and labels one part "Sam's work in 1 hour." "In 1 hour Terry could dig ¼ of the ditch." She now divides the rectangle into fourths, and labels one part "Terry's work in 1 hour." "In 1 hour they dig ¾ of the ditch. That means there's ¼ left to dig. What's left is one third as much as what they already did, so it should take one third as long, 20 minutes. That means it took 1 hour and 20 minutes to dig the whole ditch." With that, Cora turns to Jim seeking his approval.

But instead Jim turns to the class. "Hmm. Now, assuming that Terry and Sam work together until the job is done, we've got two answers."

"At least one of them is wrong," says Michael wryly.

"Which one? Or are both wrong? And why?" But before anyone can respond to Jim's questions, he glances at his watch and realizes the time. "Oh my, we have to stop."

"Well, what's the answer?" several people chorus.

Jim turns the question back. "What do you think the answer is?"

"We don't know. Aren't you going to tell us?"

"If I did, you would stop thinking about the problem."

"You just said the lesson's over."

"What I meant was that we need to end discussion of the problem for now. But you can continue thinking about it. In that sense, the lesson isn't over."

Somewhat taken aback, the class sits, perplexed, as Ellen moves to the front of the room and Jim sits down.

* * *

The first three days of the institute are largely devoted to mathematics. As has already been suggested, this offers teachers the opportunity to extend their understanding of the concepts they teach. But it also engages them in an approach to instruction, new to most of them, designed to foster the development of mathematical thinking and conceptual understanding. Jim's unwillingness to provide answers violates a pervasive assumption that closure should be reached on all problems before the end of a lesson and that teachers should explain solutions. If subjected to reflection, frustrated expectations such as these can become exemplary experiences around which the rudiments of a new practice can begin to form.

Teachers' first morning is taken up with work on a set of nonroutine problems (see Figure 2.3), but the balance of the next three days is spent exploring the properties of number systems (see Chapter 3). The first 2 hours of a typical 2½ hour period are spent on the mathematics, teachers working in small groups for about 1 hours and 15 minutes, the groups then sharing their findings with one another for the next 45 minutes. But perhaps more important for them than their investigation of any specific content area is the process of active self-reflection to which the last half hour is dedicated. By analyzing together their experience of the just-completed mathematics activity, teachers begin to construct an understanding of how knowledge develops and the circumstances that stimulate or inhibit it. This process of reflection continues in group discussion throughout the institute and is also carried on at the individual level of daily journal entries and weekly papers.

*   *   *

Although resolution of the "ditch problem" has not been achieved, staff agree that the teachers should have an opportunity to discuss their experiences of the mathematics activity before the period ends. (Those who wish to return to the problem can attend an optional session the next day.) Ellen begins the final half hour by asking, "What did you feel as you worked on this morning's problems?"

"At first it was intimidating to have to talk about my ideas out loud," says Marlene, a school administrator from Minnesota. "When an instructor came to listen, I wanted to just shut up. But after a while, as she just listened and asked questions, I stopped feeling so self-conscious. I started to see that I could solve the problem."

"I felt at times that I really needed the instructors' interventions," Betsy confesses. "As I got started on the problem, I quickly arrived at my threshold of frustration. I've never been able to do math. In fact, I hate math, and pretty soon I wanted to jump ship. When the instructors didn't arrive in time, no learning occurred, as all three of us in the group just bailed out. When the instructors were present at the moment of frustration, before total despair took over, then I

Figure 2.3 Problem Set Used on Day 1

Solve the following problems. Try at least two ways.

1. There are fewer than six dozen eggs in a basket. If I count them two at a time, there is one left over. If I count them three at a time, there are none left over. If I count them four, five, or six at a time, there are always three left over. How many eggs are there? (Weinberg & Parker, 1988)
2. It takes Sam 2 hours to dig a ditch, and it takes Terry 4 hours. If the two of them work together, and efficiently, how long will it take them to dig the ditch?
3. Emily receives her paycheck for the month. She spends 1/5 of it on food. She then spends 3/8 of what remains on her house payment. She sends 7/10 of what is now left for her other bills, and 1/3 of the remaining money for entertainment. This activity leaves her with $100. What was her original take-home pay?
4. A farmer has some hens and some rabbits in her yard. Tom sees that there are 18 heads and 50 feet among the animals. How many rabbits and how many hens does the farmer have?
5. I have trees planted in a square planting. (The number in each row equals the number in each column). I want to add more trees without losing the square configuration. The minimum number that I can add and still keep the square is 211. How many trees are in each row (just before adding the 211)?

found a small sense of pride at being considered capable of thinking in this manner. When I came to an answer I was sure of, I was positively elated!"

While Betsy is thinking about the timing of instructors' interventions, Sylvia, a third-grade teacher from Newark, is thinking about their content. "Vicki and I were discussing the importance of asking the right questions to steer people away from the wrong thing. Like, at first we solved the problem by averaging, and said it would take Sam and Terry 3 hours to dig the ditch. Then Deborah asked us how we knew to find an average. Well, we thought we had a good reason for it, but then she asked how Sam would feel about having Terry work with him since before it had taken him only 2 hours. That question made us see that our answer wasn't logical, and so we went to look for something else."

Vicki nods her head in agreement. "Her question helped us look at it in a new way. Otherwise, we would have been satisfied saying the answer was 3. But now we need to think about why averaging doesn't work in this problem."

"Well, at least we know the answer's not 3," Charlotte says impatiently. "But we're still left with whether it's 1 hour and 20 minutes or 1 hour and 22.5 minutes. Why don't they tell us?"

* * *

As instructors who work on problem solving with institute participants, we constantly ask ourselves how much information to give, which questions to answer, and which to pose. We try in our interventions to engender cognitive conflict where we see faulty logic or to stimulate further thinking where we see an opportunity to open up a new area of exploration. At times that does mean offering answers or providing additional information. But it can also mean responding with new questions, or withholding answers when teachers are capable of figuring things out for themselves. In any case, interventions are chosen not so much to help participants get right answers as to help them construct more powerful concepts.

For example, when Sylvia and Vicki attempt to solve the "ditch problem" by averaging Sam's and Terry's times, Deborah frames her questions to help them see more than the illogic of their solution, 3 hours. She also wants to expose the tenuousness of their understanding of the concept of average. For that reason she first asks them to identify which features of the problem prompted them to average. Only then does she help them see that their answer is unreasonable. Now they realize that they need to construct a more precise concept of average, one that would exclude the conditions of the "ditch problem."

Jim's interventions are likewise chosen to stimulate thinking rather than lead participants toward the correct answer or away from an incorrect one. But why not end the class session by giving the solution? Imagine that he responds to Betsy and Cora in a way more consistent with their expectations: "Look, Betsy," he might say. "See that cube that stands for the last section of the ditch they dug? You have Sam working on that section all alone while Terry's just standing around. But you started with the assumption that they worked together until the ditch was done. So, you see, Sam will do only ⅔ of that last portion while Terry does ⅓. That way it takes only 5 minutes to do that bit. Add those 5 minutes to your 1 hour and 15 minutes, and you get the same answer as Cora."

Now everyone would know the right answer, some might even have followed Jim's reasoning, most would be impressed with how smart he is, Betsy would have learned that she was wrong, and the thinking would have ended. Disequilibrium would have been diffused, and the learning process would have been truncated.

Instead, the class period ends with puzzlement unresolved. The challenge to continue thinking stands, and is met!

"What an exhausting, exhilarating day!" Betsy writes in her journal.

The A.M. session focused on solving two problems. I was completely frustrated with the ditch problem—couldn't make sense of it at all without some help from [another group. Once we finally were on our way to a solution] we struggled with what to do with the last cube, but finally convinced ourselves we were right—maybe. That hesitation left us open to discussion and, sure enough, Cora and I were discussing it repeatedly until evening,when we were both convinced she was right and I was wrong—and it was okay!

Pursuing her disagreement with Cora, Betsy not only discovers the mistaken logic in her solution, but also begins to see herself as someone able to solve mathematical problems on her own.

For many institute participants, Jim's behavior raises other questions, ones that are pedagogical in nature. They have seen Jim deliberately, and with support from the rest of the instructional staff, act in a very surprising way—a way which, they feel, contradicts some basic principles of good teaching. "When he responded neutrally to both Betsy and Cora, where was the positive feedback for the correct answer?" they ask. "How are the students going to know what's right if the teacher doesn't tell them? Isn't the teacher being irresponsible by letting a student leave class believing something that is wrong?"

As the institute models new classroom structures and new expectations for mathematics instruction, teachers must confront hitherto unexamined assumptions that form the basis of their classroom practice. Pat, for example, is disconcerted by the staff's apparent indifference to answers. It started, she thinks, with their initial instruction to the small groups to find at least *two* methods of solution to each of the problems.

> Finding the correct solution was important to me. . . . Once I felt I had a correct answer, it was difficult for me to try and think of another way to do the problem.

And then she is concerned about a whole-group discussion in which 45 minutes were spent sharing solutions to just two problems, but in which no resolution was achieved.

> If a student is giving a lengthy explanation—clearly faulty—how long should that be allowed to go on? Are there students for whom that method would be a hindrance or even harmful? Are correct answers important? What is the balance of valuing the process with real and/or accurate solutions? Because typical math classes have over-emphasized the correctness of the math we do, are we going to swing too far in the other direction to—only the process is of value?

Pat, a very thoughtful and concerned teacher, will continue thinking about these questions through the next two weeks. Whatever her conclusions, her evolving conception of effective classroom practice will be stronger for having arrived at them herself. It is important that *she* resolve these issues rather than take at face value the various teaching strategies she sees employed.

Had Pat simply been told by a lecturer that, for example, "closure at the end of a class period stifles learning," she might easily have ignored that idea or interpreted it away. Or, ludicrously, she could have understood it as a directive never again to answer any of her students' questions. But, having had her own expectations thwarted in a context that matters to her, Pat is now compelled to struggle with her disequilibrium, to consider the event with all its attendant complexity. It is on the basis of this kind of process that Pat will develop a practice that embodies a coherent, flexible, and deeply felt set of goals for student learning.

For Ana, too, being a student in the kind of mathematics class taught in this institute raises questions about her own mathematics instruction that are unlikely to have occurred to her in another setting.

> I learned a lot from the experience. The group made me think. We wanted to find a solution to the problem. We had a lot of time and we could find a solution to the problem. It made me think about the time I allow students to solve problems. . . . It takes time for children to go through the processes of problem solving.

And as Betsy thinks about a theoretical question raised in one of the assigned readings, these first three days of mathematics explorations provide her reflection with an experiential basis:

> I read the article [Peterson, Fennema, & Carpenter, 1989] and was disconcerted by a number of things. . . . They say lower-order computation is not prerequisite to higher-order problem solving, but rather that they are learned together. . . . No longer may we teach from the simple to the complex. The simple is mastered as the complex is learned. But wait—wouldn't that be confusing? No—I mastered that Xmania lower-order counting by creating a higher-order place-value system. Hm. . . . Similarly, I continued to work on computation proficiency as I worked with place-value charts. Well.

For all three teachers, it is the insight into the learning process yielded by their own self-observation that allows them to see issues so clearly. Later in the week, Ana addresses just this aspect of the institute experience:

> [I] began to think that I have attended a lot of workshops in [alternative instructional approaches for] math . . . and that I believe in the theory. Though my classroom has not changed much. Why? It goes back to the reliance on what I know, the way I was taught.

What is so special about this program is that I have experienced as the students experience in the classroom. I would never forget the experiences. [I] experience a lot of good feelings [as well as] frustration, but at the end a sense of accomplishment. Very good way to teach others, by experiencing first.

LEARNING TO LISTEN FOR STUDENTS' UNDERSTANDINGS

The new instructional paradigm requires that teachers learn to ask the kinds of questions that will enable them to identify their students' mathematical conceptions and the issues with which they are grappling. But in order to know what to ask, teachers must learn how to listen—what to listen for—in what their students are saying. To this end, after three days of active verbalized problem solving, institute participants concentrate on listening. Working with videotapes and in live, one-on-one interviews, they are asked to analyze students' solution processes, assessing the extent of their understanding and exploring the significance of the gaps that are exposed.

Teachers tend both to under- and to overestimate children's mathematical understanding. On the one hand, they do not believe that their students can figure something out on their own, can know something before it is covered in class. But on the other hand, they must come to appreciate that severe misunderstandings can lie behind correct answers.

Among the tapes the teachers view, two vividly illustrate these points. On the first, two 10-year-olds, Jennifer and Eliza, are at work on the "egg problem" (see Figure 2.3), which the teachers themselves had been given on the first day of the institute to ensure they would not be distracted from following the girls' reasoning. The teachers witness how, after an initial and frustrating attempt at solution by trial and error, the girls devise a more methodical approach and soon begin to see patterns in the possible answers they eliminate—for example, that any number which yields a remainder of 3 when divided by 5 must end in 3 or 8, a surprisingly sophisticated result that the teachers themselves had struggled over. After working on the problem for nearly 20 minutes, Jennifer and Eliza delightedly conclude that 63 satisfies all the conditions of the problem. These two 10-year-olds instance some impressively powerful mathematical thinking.

As to the second point—that severe misunderstandings may comfortably coexist with the right answer—many teachers believe that if material has been covered in class and their students can reproduce their teachers' solution processes, there is no more to be said. The second tape segment (Ginsburg & Kaplan, 1988) suggests otherwise:

"What does this little line up here mean?" asks Rochelle Kaplan, a developmental psychologist, of Andrea, a 7-year-old who has just shown her that 15 + 7 = 22.

"It's a *1*. See, 5 and 7 is 12, so I put down the *2* and carry the *1*."

"Why don't you put the *1* down here, next to the *2*, so it says *12*?"

Andrea looks at Rochelle curiously and then starts to explain. "No! You see, this is the ones column and this is the tens column. You can't put the *1* down there because it doesn't fit in the ones column. You have to put it up here. And then you go 1 + 1 = 2, and then you put the *2* down here, *22*."

Andrea is very confident, but Rochelle continues to probe. She writes the number *17*. "The *7* is in the ones column and the *1* is in the tens column," Andrea declares.

Now Rochelle points to a pile of chips on the table. "Can you show me 17 with those?"

"Sure." Andrea counts out 17 chips, then counts again to make sure she hasn't made a mistake.

"Now show me how much the *7* here in *17* stands for."

Andrea counts out the right number of chips. "Seven," she shrugs.

"And show me how much the *1* in the *17* stands for."

Andrea now pulls out 1 chip. "One," she says, looking up at Rochelle with a smile.

"Well, what about all of these?" asks Rochelle pointing to the 9 chips that had been unaccounted for.

"Those are extra."

*   *   *

"Watching the videos," Ana later writes,

> I began to ask myself, what is learning? Especially because Andrea did what all children in my classroom would have done. That's the way I teach place value, and the children don't have any time to explore with manipulatives. It reinforces the idea of the importance of the manipulatives in the classroom. . . . [And I see] the importance of really observing the children, asking questions, to see if they really understand what they are doing.

While the mathematics activities that began the institute introduced Ana to new ways of organizing the classroom and to new instructional strategies, it is Andrea who really causes her to begin seriously to reevaluate the core assumptions of her practice.

> I'm not quite sure [what] my philosophies about learning are. I'm beginning to question my teaching strategies. What the nature of the learning environment should be, and what kind of learning experiences [I should provide], etc.

While Pat has found reconstructing Eliza's and Jennifer's solution process to be a challenge, for her, too, it is Andrea who keeps coming back as she writes in her journal that evening. But now she questions her initial assessment of Andrea's understanding:

> I [first] assumed that Andrea had no idea what a 1 in the tens place was . . . [and it] may indeed be a lack of understanding of the value of numerals in the various positions. [But] it may be a lack of understanding of how to use the manipulatives to represent an amount.

Pat realizes that, just as a correct answer does not in itself certify understanding, one incorrect response does not unambiguously establish what the child's conception is. "More questions," she writes, "would have been very much needed. . . . The questioning may have been the confusing part to the child. We can think we are asking one thing and really be saying something completely different!"

One-on-one interviews with local children provide teachers with just such opportunities to probe mathematical thinking in as much depth as they feel is needed. These children are recruited for two days (Spanish-speaking children are recruited for those teachers who are more comfortable using Spanish) and each institute participant is assigned a young partner. The objective of the first day's work is for each teacher to obtain a picture of what her partner understands and what he or she finds confusing. Based on this information, the teacher decides that evening what her partner might usefully learn next and, with new instructional strategies in mind, designs an appropriate lesson. Next day, teacher and child try it out.

For her second day's session, Pat carefully designs a set of word problems addressing a particular aspect of division, but she later realizes that it was the questions spontaneously elicited by her partner's solutions that helped bring about the understanding she was after. "I didn't ever intend that the two problems would 'do the trick' but it just stood out to me how important questioning and guided thinking is. . . . I also learned that in no way can I anticipate all the right questions ahead of time."

Betsy's experience of this exercise is also positive, but more complicated. She discovers that listening to her students reason mathematically can threaten her own sense of competence. Several days before the live interview, she had found herself demoralized by Eliza's and Jennifer's success in solving the "egg problem."

> How come no one else mentioned how embarrassing it is to hear the fluency in thinking of the girls and comparing it to their own? . . . I'll never be able to teach like this. I guess I'm in total disequilibrium though I've spent much time denying it.

But now she has the opportunity to work with a live, mathematically in-
clined sixth grader, and she is intrigued.

Michael challenges me to think! . . . We began comparing fractions. It
was simple [for him] to represent until 4/9 and 4/10. No manipulatives had
ninths, so what to do? He tried 9 cubes, removing 4, [and] compared
[that] to 10 cubes, removing 4. Seeing 4 in each pile, he said, "That
doesn't prove anything." So he looked at his remainder, saw 6/10 and 5/9 and
said [while pointing to 6/10], "Since the denominator is larger, this is the
smaller piece." I asked him to show me another way. . . . He drew pies,
divided one into ninths, one into tenths. Then he said, "This one [10th]
turns 36 deg[rees] each, that one 40 deg[rees] each, so that one
[4/9] is bigger." I was in total awe. I asked for more explanation and he
multiplied 4 × 36 and 4 × 40 and again demonstrated which was bigger.

Rather than feeling intimidated by this child's mathematical thinking,
which is, she feels, superior to her own, Betsy learns that by being prepared to
learn from him she can help him, too. This discovery not only transforms her
conception of what should take place in a mathematics classroom, it also
strongly affects her conception of herself and what she can do.

I can teach even the brightest, even though I don't know all he knows. I
can work with him and learn from him 'til we can both agree on an answer. I learned how incredibly rich and creative thinking can be if it's not
locked into a [single] answer. I felt freed to think again with him, not
driven to stay ahead of him, which was an impossible task given our individual abilities. I learned *not* to get in his way.

TEACHING TO THE BIG IDEAS

Many teachers who come to these institutes complain of the pressure they
feel to cover the material specified in the detailed curricula they are handed at
the start of the school year. Most elementary teachers also admit to never having
overcome their own early experience of mathematics as a bewildering succession
of math facts and computational procedures anxiously committed to memory.
From the standpoint of an instructional paradigm informed by constructivist
commitments, the pressure to move quickly and the lack of understanding of
mathematical principle are related phenomena. If teachers don't know what to
teach for, covering all the material seems to provide reassurance that nothing of
importance will be left out. In fact the opposite is closer to the truth: the conventional topic-by-topic, drill-and-practice pattern of mathematics instruction

subverts student understanding of the principles that underlie mathematical order. The SummerMath for Teachers staff has begun to refer to these principles as "big ideas," and their thinking about mathematics curriculum increasingly revolves around the concept of teaching for the construction of big ideas.

Big ideas are the central, organizing ideas of mathematics—principles that define mathematical order. One such principle, encountered early in one's mathematical development, is that just as individual objects can be counted, so, too, can some number of objects, 10, say, be counted as one group. In other words, one can count groups of objects as well as the discrete units that comprise them. As a matter of logic, every counting system possesses this property; as a matter of convention, some counting systems, place-value systems, employ the same numeral to denote an individual, or a group of individuals, or a group of groups of individuals, and so on, depending on the numeral's position in a numerical expression. Thus, in the base-ten place-value system, in the expression *33*, the *3* furthest to the right denotes 3 ones, while the *3* to the left of it denotes 3 groups of 10 ones. The idea that a numeral can refer to either groups or units is encountered again when children learn that 10 units can be "regrouped" into 1 ten or vice versa. Later still, the same big idea is encountered in the operation of multiplication where, for instance, the expression *3 × 4* indicates 3 groups of 4 ones.

A second example of a big idea is that addition and subtraction are inverse operations. Children initially understand these two operations as distinct and "opposite" actions—"adding on" or "taking away." Hence, the difficulty they have conceding that some problems are solvable using either operation—for example, *I have 24 pencils. I give one to each of the 18 children in the class. How many am I left with?* Here, the answer, 6, can be arrived at either by "taking away" 18 from 24 or "adding on" 6 ones to 18 to reach 24. This same principle arises in the analogous relationship of multiplication and division, and grasping it is an important prerequisite for understanding equivalent equations.

Still a third big idea, this time drawn from upper elementary mathematics, and one with which many students wrestle, is that the quantity represented by a fraction depends on its reference whole, so that a single quantity may be represented by different fractions, depending on the whole to which it refers. For example, consider this context for $\frac{2}{3} \times \frac{3}{4} = \frac{1}{2}$: *Jody had ¾ of a yard of fabric and used ⅔ of it to make an apron. How much of a yard was needed to make the apron?* In this case, the amount of fabric used to make the apron is identified both as ½ and as ⅔—½ of a yard, but ⅔ of Jody's fabric.

Teaching to the big ideas means facilitating their construction by providing contexts for relevant mathematical explorations. Typically, developing a big idea entails seeing old strategies in a new way: formerly unrelated rules and results are now understood to share an underlying unity. But, to echo a point made in Chapter 1, some big ideas are never fully and finally comprehended. Rather, their construction is always provisional—they will be successively reorganized, acquiring additional content, as new areas of mathematics are explored.

The concept of big ideas is broached early in the institute as participants are asked to begin thinking about the three to five most important mathematical ideas they want their students to learn by the end of the school year. But when, on the second-to-last day, the teachers are finally invited to discuss their conclusions, this exercise frequently proves too sophisticated for their level of development. For most, their own mathematical understanding is simply not rich enough to allow them to think in terms of "ideas" rather than "topics." Thus, they usually interpret the task to mean that they should consider the topics they are charged with teaching, setting priorities to determine the three to five most important items on their list.

While the institute staff's ultimate goal in designing this activity is a radically reconceived mathematics curriculum, the task the teachers define for themselves is also significant. The unfamiliar notion that they can take control of the content of their instruction provides some teachers with the encouragement they need to begin to implement new approaches to teaching. Pat registers this insight:

> As the teacher, I need to make serious decisions about what ought to be taught and ways to challenge and increase my students' understanding. As I become an observer of my students while they work to solve problems both alone and in the group setting, and gather information about how students make connections and add to their understanding, I can then make informed decisions about the content and pace of the curriculum. The emphasis is off the quantity and on the quality of learning and instruction.

As they engage in this activity, it turns out that most participants have not been able to see past the familiar topical organization of the mathematics curriculum, but eventually the whole group discussion does turn to the question, What is a big idea? Here a seed is sown. And as teachers go through the year reengaging familiar curricular material, they may begin to evaluate the content they teach in the light of the distinction between topics and big ideas. As Jill Lester's story, told in Chapter 5, illustrates, often teachers themselves come to understand what the big ideas are to which they should be teaching only as they work to transform their practice. Those who return to an advanced institute will be given the assignment again.

NEW VISIONS

The institutes' activities are designed to stimulate the imagination—as teachers gain insight into the learning process and are exposed to novel ways of thinking about mathematics, they are also constructing the outlines of an alternative mathematics instruction.

Before describing her new sense of the kind of teacher she can be, Ana assesses her pre-institute classroom practice:

> My . . . students used to sit in my classroom very passively, listening to me explain problems and how to solve them. The students never did much questioning and were never encouraged to verbalize their mathematical ideas. The students were never really excited about doing math. Students in my class never realized that some questions had more than one answer. There was not time for reasoning or solving a problem without my help. I had my students memorize examples such as addition tables instead of having them seek independent solutions. . . . Workbooks [and] ditto sheets dominated my teaching lessons. . . . The math lessons were done in a large group. Children work[ed] individually at desks, most of the time listen[ing] to my directions in a whole group.

Through her experiences in the institute, Ana decides that her classroom lacks student engagement—there is little excitement or independent thinking. The appeal of these qualities evokes for her a picture of what her classroom could become.

> Now I see myself as a facilitator. . . . Now I understand how children work better cooperatively, in small groups. Children should be provided many opportunities to develop social skills such as cooperating, helping, negotiating and talking with the other person involved in solving the problem. [I must] accept and recognize that often there is more than one answer to a problem. . . . The quality of learning that takes place from self-directed problem solving and experimentation is important. The curiosity of people is a powerful teacher. . . . If the environment that we set up for learning is meaningful to them, they will be motivated enough so that they will try to create meaning out of whatever it is that we are presenting to them.

Betsy writes about how the institute has transformed her understanding of the nature of mathematics.

> Today I found myself thinking quite differently about math and would like to share some of the notes I jotted down. . . .
> a way of *thinking*, not rote memorization or meaningless computation
> a mirror, reflecting one's willingness to wrestle with the tough questions of "Why is that so?" or "How can I prove that?"—perhaps even reflecting where one is on the developmental ladder of critical thinking

a way of loving students, enabling them to make their own discoveries,
respecting each child wherever s/he is, affirming, gently challenging,
supporting through the disequilibrium
a way of relating without judgment
a way of being in the world—listening, communicating, wrestling with
big ideas, reflecting, discussing
a way of meaning-making.

But in the face of decades of traditional mathematics instruction, Betsy
feels unsure of her ability to create a classroom where all this can be realized.
Nonetheless, she finds in these glimpsed possibilities a vision of a classroom in
which she and her students engage together in this new kind of mathematics.
And she now understands that if this is to happen, she must learn to listen to her
students' mathematical ideas in a new way.

> [I am] no longer . . . the source of all knowledge, the judge of right an-
> swers, the dispenser of good/bad grades. Instead, I must learn to under-
> stand how the children are thinking about the problem. . . . I must de-
> velop a greatly improved style of questioning, where I look not to lead
> them to the right answer, but where I truly understand their method of
> working on the problem. . . . As we meet together as a class, they will
> have the opportunity to further reinforce their learnings by articulating
> them clearly in the group and answering questions from other classmates
> who are struggling to understand another way of solving the problem.
> These techniques surely represent a major change in my thinking about
> teaching; in fact, it feels cataclysmic!

Betsy has become committed to involving her students in the kind of mathematics she encountered in the institute. From the types of classroom instruction
she has experienced and the listening activities she has engaged in, she selects
strategies she believes will help her realize her new goals.

Pat has previously studied Piagetian theory, but she has not been sure of its
implication for her mathematics instruction. During the institute she has developed a clearer vision:

> What has changed significantly for me is that I now see more clearly that
> my previous model of teaching . . . did not match my belief about how
> kids learn. I was aware that math was not like the other instruction going
> on in my room, but [it was] unclear what the problem really was. I did not
> provide the opportunity for my students to invent their own way to under-
> stand the concepts and content for themselves. They were supposed to
> come to my understanding or a textbook writer's understanding of math.

And though she does not now feel that she has all the answers, Pat has some clear ideas about directions for her own continued development.

There are no simple solutions and I certainly don't feel like I've been exposed to more than the tip of the iceberg, but even that tip has given me a wealth of assistance in rethinking what needs to begin to take place in my classroom. Learning by discovery is not new, but I think misunderstood. If I am to [create] a classroom where students can discover and make sense of mathematics for themselves, I need to be very clear in my own mind what are the important concepts and content areas that I am going to try and help them to "discover." . . . There is a lot of mathematical thinking I need to do in order for me to create the situations that will facilitate my students' discovery of math concepts. Otherwise I'm just leaving it to chance. As I ask myself those kinds of questions that challenge my own understanding of math, I will be better equipped to create methods that will enhance learning. . . .

My role as the teacher is . . . to determine how the learning environment in my class will help the students build the bridges connecting concepts and knowledge previously understood to new levels of understanding. That involves so much more than simply giving a book pretest. It seems the more I reflect on what my teaching model is, the more I realize I must study and talk to the learner in order to gain a deeper understanding of his or her way of learning.

"I think I can do it," Pat writes at the end of the institute.

We studied the *NCTM Standards* during our curriculum meetings last year. They frightened me. I believed what they said about math in our schools today. I knew that I was guilty of some inferior teaching practices and I felt somewhat powerless to do anything about that. These two weeks have given me a sense of empowerment. I'm no longer afraid. I have a long way to go, but I have a sense of where I'm going.

Cultivating visions of an instructional practice consistent with the new paradigm necessarily calls into question many habits and assumptions about mathematics and about learning and teaching. Inevitably, this entails a more or less protracted period of disequilibrium. And because teachers' conceptions of teaching are so closely tied to their professional identities, the emotions aroused are intense indeed. At the same time, it is the teachers' own internally generated need to resolve this unbalancing that creates openings for positive change. Initially, most institute participants feel anger, frustration, or even despair; but for many, these will give way to excitement and even elation as the institute progresses.

Teachers leave the institute with unfinished, in some respects crude, yet, for many, compelling visions of a new kind of mathematics classroom. They will need the few weeks remaining before school starts up again to "decompress," to relax, reflect, and prepare for the coming year. But as they reenter their classrooms, animated by fresh "images of the possible and desirable," they find the sheer mass of familiar daily realities pulling them back into the old orbits of safe routine. For most teachers, then, whether their "commitments to make those images a reality" will be sustained will depend on the availability of similarly committed follow-up support.

3
A Sample Mathematics Lesson for Teachers: Xmania

How can teachers teach a mathematics that they never learned, in ways that they never experienced?

David K. Cohen and Deborah Loewenberg Ball,
"Policy and Practice: An Overview"

Chapter 2 described how SummerMath for Teachers uses mathematics activities to help teachers envision new possibilities for their own classroom practice. In this chapter, the focus is on the mathematics itself. Xmania is a lesson intended both to strengthen mathematical understanding and to suggest an expanded view of the nature of the discipline. More specifically, the activity is designed to stimulate new ideas about number systems in general and place value in particular. It is structured so as to demonstrate to participants the extent to which the mathematics they do is their own creation.

THE ASSIGNMENT

Thirty-six teachers seat themselves at tables in threes and fours. They know that the agenda for the next two class sessions is "Xmania: An activity about number systems," and they are curious about it. The mathematics lesson begins with Deborah telling the following story:

"There is a civilization far, far away called Xmania. In fact, it is so far away it has no contact with our civilization, except for this one story that I heard. On the planet of Xmania the people had a number system that went like this." Deborah holds up her fist: "*0.*" Then she holds up one finger: "*A,*" two fingers: "*B,*" then an additional finger each for "*C, D, E, F, G, H, I, J,*" finishing by flashing both hands twice and then six fingers, "all the way to Z. Anything more than Z was *many*.

"Actually the people of Xmania were very dissatisfied with their own number system."

"I can see why! It wasn't specific enough," suggests Violet. "Different numbers are represented by the same word, *many.*"

"That's a problem, isn't it? $Z + A$ is many. $Z + Z$ is many. And also $Z \times Z$ is many. In fact, the Xmanians were so dissatisfied they gave all their mathematicians generous grants to work on designing a different number system.

"Xmania's mathematicians worked on this problem for several years until one day a certain mathematician announced the successful design of a new system. This new system, the mathematician said, uses only the symbols *0, A, B, C,* and *D*; in this system you can represent any number, no matter how large; and you can add, subtract, multiply, and divide.

"The only problem was that before the mathematician was able to publish a full description of the number system, she died. But she left behind these objects that she was going to use to explain it." Deborah holds up a sample of each of the objects shown in Figure 3.1.

"And so, your task is, in each of your groups, to come up with a possible number system for Xmania that fits the description. That is, it is based only on the symbols *0, A, B, C,* and *D*; it can represent any number; and you can operate in it. And you should use these blocks to help you develop and explain the system."

(In order to understand the processes of the teachers, you, the reader, are encouraged to explore Xmania for yourself. Figure 3.2 displays a page of two-dimensional units, rods, and flats that you can copy, cut out, and use to experiment with the assignment. You will find it helpful to use these figures as you follow participants' logic. If possible, find a partner with whom you can explore

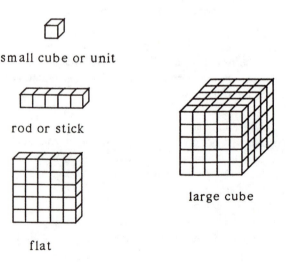

small cube or unit

rod or stick

large cube

flat

FIGURE 3.1. Xmania blocks

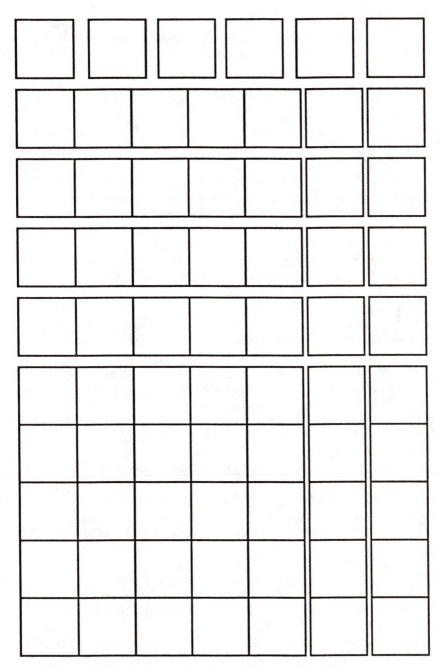

FIGURE 3.2. Two-dimensional Xmania blocks to copy and cut out

the mathematics. Following the mathematical arguments requires close attention. If you do not have time to read slowly, working through the mathematical
reasoning, skim this chapter now and return to it later.)

INITIAL EXPLORATIONS

Like most mathematical activities in SummerMath for Teachers institutes
and courses, the lesson on the Xmanian number system begins with small-group
exploration. This instructional technique serves a number of important purposes.
First, there is an immediate shift of attention, and authority, away from the
instructor: responsibility for the group's work comes instead to reside in its
members. This shift makes it clear that work will not get done by following
step-by-step procedures dictated by someone else—that participants will need
to generate their own interpretations, conjectures, and solution strategies. And
in so doing, they will begin to experience mathematics as something human
beings do—in this case, as their own construction. Second, as participants communicate their ideas to one another, they achieve greater clarity for themselves.
Teachers frequently remark that they come to understand the content they teach
when they have to explain it. Here that process is generalized. Finally, within
the constraints of the assignment, students address the mathematics at their own
pace. As each of the groups works, it necessarily focuses on those aspects of the
activity that its members have the most difficulty understanding.

This morning's instructional team—Deborah, Jim, and Ellen—distributes
the blocks as the groups start in on the problem. When staff are assured that the
groups are functioning, they give them a few minutes to get started and then
begin to visit, listening as teachers reason aloud, asking questions where appropriate.

"Can you explain your system to me?" Deborah asks of the first group she
visits.

The group's members glance at one another to decide who will respond.
Carmen speaks up confidently. "Well, *0* stands for zero, nothing," she says.
"Then we have the other four symbols: *A, B, C,* and *D.* We also have four
objects to work with. So we'll call the small cube *A,* the rod *B,* and the flat *C,*
and the big cube *D.* Then we can count: *A, AA, AAA, AAAA, B, BA, BAA,
BAAA.* We have to keep going to see what happens with larger numbers."

Carmen's group has created a system that resembles the ancient Egyptians'.
While they assigned symbols to powers of ten (1, 10, 100, 1,000, etc.), this
group assigns them to powers of five (1, 5, 25, 125). (See Figure 3.3.) Deborah
leaves them to work on their system further and moves on to a second group.

At Ann's table, she finds the chart displayed in Figure 3.4 and notices that
the value "one cube" has been assigned to the symbol *0.*

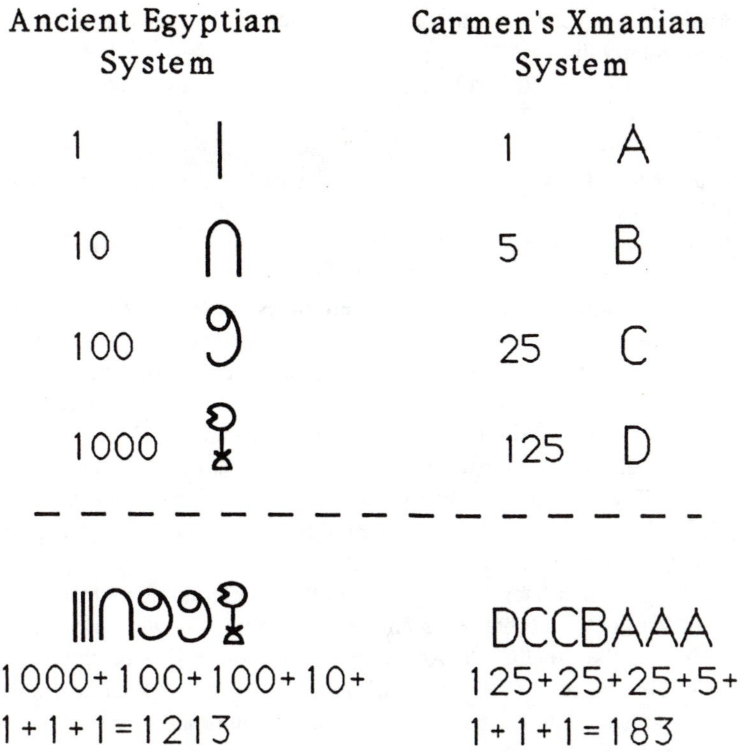

Ancient Egyptian System		Carmen's Xmanian System	
1	|	1	A
10	∩	5	B
100	℃	25	C
1000	⚷	125	D

‖∩℃℃⚷

1000+100+100+10+
1+1+1=1213

DCCBAAA

125+25+25+5+
1+1+1=183

FIGURE 3.3. Comparison between Carmen's system and the
Ancient Egyptian System

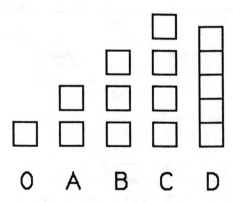

O   A   B   C   D

FIGURE 3.4. Ann's group's first assignment of quantity to symbol

Ann taps her pencil on the table impatiently. "Now what comes after *D*? What comes after the stick?"

"You add a cube," Robert responds. "It's made like this, a stick and a small cube."

"Yeah, what are we going to call it?"

"*D0*," which she pronounces *dee-oh*. "It's made of a stick and a cube, a *D* and an *0*."

"Okay, so we can go on. A stick and two cubes is *DA*. Then *DB, DC,* and *DD*."

Up to this point, their system also resembles some of the ancient number systems. In the Hebrew and Greek systems, for example, the first ten numbers are represented by the first ten letters of the alphabet. Eleven is represented by the symbols for ten and one written in a string, twelve by the symbols for ten and two, and so on. Nor do these systems possess a concept of zero.

Nancy looks at the *DD* uncertainly. "But look at the flat. It's made of *D D*'s. If we call two sticks *DD*, then what are we going to call a flat that is made up of *D* sticks? See, *0, A, B, C, D.*" She holds a rod against a flat and counts to show how many rods make up a flat.

"We'll get to that later. Let's work with the smaller numbers first."

"Wait a minute." Robert sees another problem. "Do we really want *0* to represent one cube? In the old Xmanian system, she had *A* be this," he says holding up one finger, "and *0* was this," holding up a fist. "What do we want *0* to be? Do we need *0* to be zero? It seems pretty important to have *0* stand for nothing, zero."

Immediately convinced, Ann rearranges the cubes on her chart. "All right. If *0* is zero, we have *A* is one cube, *B* is two cubes, *C* is three cubes, and *D* is four cubes. Well then, what is the symbol for the stick?"

Assured that this group is finding its own questions to pose and explore, Deborah moves on to Lyanne's group. They, too, have assigned *A, B, C,* and *D* to the first four counting numbers, and have answered for themselves Ann's question about what comes next. Lyanne stands over their counting chart, arranging blocks to demonstrate the value of each number:

A,   B,   C,   D,   D0,
D0A, D0B, D0C, D0D, ____

"How did you decide on this?" Deborah points to the *D0*.

"You know, it's like ten. Ten ends in *0*, so this can, too," says Lyanne.

"And how about this?" Deborah points to *D0A*.

"It's made of a rod, *D0*, and a cube, *A*. So together it's *D0A*."

Deborah looks around at the other group members questioningly. Victor speaks up. "Sure. It's like a six. Five and one. *D0* and *A*. So it's *D0A*."

Here the group is mimicking the familiar base-ten number system by using the *0* as a digit in a numeral—however, without introducing place value, since *D0* is treated as a single symbol. The group does not operate on the two symbols *D* and *0* to determine the value of *D0*. Instead, they get the next number by simply concatenating *D0* and *A*. This is similar to what many young children do who see *100* as a single symbol and write its successor as *1001*.

"Our question now," continues Lyanne, "is what to do after *D0D*," pointing to the space designated by two rods, following one rod (*D0*) and four cubes (*D*).

"Maybe it should be *D0D0*," suggests Cassie hesitantly.

Victor agrees. "*D0D0* makes sense. After all, it's two rods, two *D0*'s."

"But you don't say 'ten ten,' you say 'twenty,'" says Lyanne. "You have to come up with a new name."

"Oh, right."

"So let's call it *D00*. After that, then we have *D00A, D00B, D00C, D00D*. Now what's next?"

"*D0D0D0?*"

"It's the same thing," Lyanne retorts impatiently. "We don't say 'ten ten ten.' Let's try *D000*."

In each of the systems these groups have thus far designed, there are two ways to assign a numeral, a set of symbols, to a given quantity: (a) make up a new name or symbol for the value, or (b) create a string of symbols whose values add up to the given quantity. In Carmen's system the symbols *A, B, C,* and *D* are assigned to the quantities 1, 5, 25, 125, and the value of any number—for example, *DBBA*—is determined by summing, thus, 125 + 5 + 5 + 1. In Lyanne's system, the symbols, *A, B, C,* and *D* are assigned to the values 1, 2, 3, and 4. But, even as she tries to imitate the base-ten place-value system, she ignores the place-value mechanism and assigns the symbols *D0* or *D00* as new names for given quantities, treating them just like *A, B, C,* or *D*. So, for example, to create the number that followed *D0*, she attaches an *A* at the end of the string to form *D0A*, whose value is *D0* + *A*.

In contrast to these systems, some groups introduce into theirs an analog of the Hindu-Arabic place-value system commonly used. They begin with the following values: *0*—zero, *A*—one cube, *B*—two cubes, *C*—three cubes, *D*—four cubes. The next number, represented by a rod—*A* rod and *0* cubes—is written as *A0*. A rod and *A* cube is *AA*. Adding a cube each time, you get *AB, AC, AD*, until five cubes are traded for a rod, and you have *B* rods and *0* cubes—*B0*. Then *BA, BB, BC, BD, C0, CA*, and so on. In this system, a base-five place-value system, the *C* in *CA*, for example, does not have the same value as the *C* in *AC*, or as *C* when standing alone. *C*'s position in the string determines the value it represents: in *CA* it represents *C* rods, while in *AC* it represents *C* cubes. Another way of looking at it is to say that not only does each symbol have a value, but each place has a value, too. So *the value of a number is determined by multiplying the value of each symbol by the value of the place that it occupies and then summing.*

For most teachers, adapting the essential features of place value to the constraints of Xmania is not an obvious strategy. They must suggest a system, explore it, encounter its limitations, and redesign it to overcome them. Through this process, they discover the various properties of different number systems and gain deeper understanding of our own.

Although it is intended that everyone eventually explore a base-five place-value system, the purpose of the activity is not to get each group there as quickly as possible. In fact, those groups constructing other systems are doing indispensable work. Through creating alternatives and confronting their limitations, the power of the idea of place value becomes clearer to all. Furthermore, many teachers are designing systems that closely resemble the constructions (frequently considered misconceptions) of their own elementary-grade students. Recognizing that will yield valuable insights into the children's reasoning processes.

SHARING SYSTEMS

While the work done in each group is important, much is also gained as the groups come together to share their discoveries and inventions. Specifically, in the Xmania exercise, such sharing allows participants to look at the variety of possible systems, to analyze them for their strengths and limitations, and to consider which properties systems share and where they differ.

Furthermore, when participants are responsible for presenting their results to the entire class, they realize that their ideas are important. Indeed, as those ideas are taken seriously by fellow students and by staff, the ideas' proponents learn to take them seriously, too.

Whole-group discussion also reinforces some of the lessons of small-group work—particularly, the idea that mathematics is a human construction, historically conditioned, part necessity and part convention, and open-ended.

It is with these considerations in mind that staff ask each group to design posters illustrating the number system it has created. Thus, as participants and staff enter the classroom for the second Xmania session, they find these posters displayed on the walls, and the next 15 minutes are spent touring the room studying them before the sharing starts.

Ellen leads the discussion. Recognizing that many groups had started with a system similar to Carmen's, Ellen asks Carmen's group to present first. Carmen herself comes up to the front of the room and stands by her poster.

"*A* is worth a small cube, *B* a rod, *C* a flat, and *D* a large cube." (See Figure

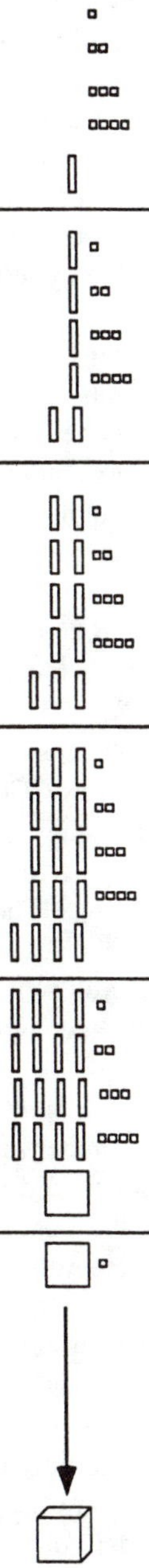

Old Xmanian System	Representation with Blocks	Carmen's System	Ann's System	Sharon's System	Julie's System
A	▫	A	A	Aa	A
B	▫▫	AA	B	Ab	B
C	▫▫▫	AAA	C	Ac	C
D	▫▫▫▫	AAAA	D	Ad	D
E	▯	B	I (stick)	Ba0	A0
F	▯ ▫	BA	I A	BaAa	AA
G	▯ ▫▫	BAA	I B	BaAb	AB
H	▯ ▫▫▫	BAAA	I C	BaAc	AC
I	▯ ▫▫▫▫	BAAAA	I D	BaAd	AD
J	▯ ▯	BB	II	Bb0	B0
K	▯ ▯ ▫	BBA	II A	BbAa	BA
L	▯ ▯ ▫▫	BBAA	II B	BbAb	BB
M	▯ ▯ ▫▫▫	BBAAA	II C	BbAc	BC
N	▯ ▯ ▫▫▫▫	BBAAAA	II D	BbAd	BD
O	▯ ▯ ▯	BBB	III	Bc0	C0
P	▯ ▯ ▯ ▫	BBBA	III A	BcAa	CA
Q	▯ ▯ ▯ ▫▫	BBBAA	III B	BcAb	CB
R	▯ ▯ ▯ ▫▫▫	BBBAAA	III C	BcAc	CC
S	▯ ▯ ▯ ▫▫▫▫	BBBAAAA	III D	BcAd	CD
T	▯ ▯ ▯ ▯	BBBB	IIII	Bd0	D0
U	▯ ▯ ▯ ▯ ▫	BBBBA	IIII A	BdAa	DA
V	▯ ▯ ▯ ▯ ▫▫	BBBBAA	IIII B	BdAb	DB
W	▯ ▯ ▯ ▯ ▫▫▫	BBBBAAA	IIII C	BdAc	DC
X	▯ ▯ ▯ ▯ ▫▫▫▫	BBBBAAAA	IIII D	BdAd	DD
Y	☐	C	d (doozie)	Ca00	A00
Z	☐ ▫	CA	d A	Ca0Aa	A0A
Many ↓	↓ (cube)	↓ D	↓ db (doozie box)	↓ Da000	↓ A000

FIGURE 3.5. Comparison of Xmanian systems

3.5.) "And this is how we count in the system. We start with *A, AA, AAA, AAAA, B*. After *B* we go *BA, BAA, BAAA, BAAAA, BB*, and so on. You never end up with more than four of the same letter in a string."

As Carmen finishes her initial explanation, the floor is opened to questions or comments.

"When I look at your chart, there's something I don't understand," says Tammy. "In the old system, we said the problem was that you couldn't distinguish among any of the numbers larger than *Z*. Well, here you're just renaming *many* and calling it *D*."

"Oh, no, that's not what we mean. You can have *CAA*—a flat and two units—and after that comes *CAAA*—a flat and three units. That's different from *CC*—two flats—or from *D*—a large cube. In the old system, all those amounts are called *many*. *D* in the new system stands for a large number that was not distinguishable from other large numbers in the old system. It would have been called *many* along with lots of other numbers. Now *D* is specifically the number of small cubes in the large cube. That's 125.

"To add," she continues, "for example, if you take *BAAA* + *CBAAA*, you'd get *CBBAAAAAA*. But that's too many *A*'s, so you make five of them into a *B* and you end up with *CBBBA*."

"The system seems to work," Roger observes, "but it's pretty unwieldy. Suppose, say, you take the number 79. You'd write that as *CCCAAAA*. It takes up seven spaces. When you get into hundreds and thousands you'll have really long strings."

A number of teachers nod their heads in agreement. "That is a problem, isn't it?" Carmen acknowledges.

"I have another question," Marsha says. "How do you write 147?"

Everybody in the class works on that problem and agrees that 147 is *DBBBBAA*.

"Okay, now does the order of the letters make a difference in this system? What if I wrote *ABDBBAB*? What is the value of that number in this system?"

The class mulls this over and some teachers start to add up the values of each of the symbols in *ABDBBAB*. Carmen, up at the board, stops midway through her calculation. "Oh, if you're just adding the values, order doesn't matter."

Ellen lets the class ponder this feature of Carmen's system before suggesting that they move on to Ann's group.

"Here's the basic pattern," Ann says. "*0* is zero. *A* is this," holding up one finger; "*B* is this," holding up two fingers; "*C* is this," holding up three fingers; "and *D* is this," holding up four fingers. "We called the rod *stick*. Then, you see, we really had fun making up names—we called the flat *doozie*. When we had this many sticks," holding up five fingers, "we had a doozie. And this many doozies," again holding up five fingers, "is a *doozie box*. So we can count: *A*,

B, C, D, stick, stick-A, stick-B, stick-C, stick-D, stick-stick, stick-stick-A, and it goes on." (See Figure 3.5.)

Ellen asks, "First, does anybody have questions about how the system works?" She waits, but there are none, so she asks one of her own. "How is this system the same or different from the other one we just looked at?"

"This one added more symbols. They used more than just *0, A, B, C,* and *D.*" Ann nods her head in acknowledgment.

"Order doesn't matter in this system, just like the other. You can have *doozie-stick-stick-D* or *stick-D-stick-doozie.* They both stand for the same amount."

"No," Ann disagrees. "That's not what we had in mind. We can just decide what order to put things in. It needs to be consistent, but we can say that the symbols with the larger values go on the left. *Stick-D-stick-doozie* is simply not a correct way to write a number in our system. You know, like not every string of letters is a word—'c-a-t' spells *cat,* but 't-c-a' doesn't spell anything."

In this way Ann resolves the question that was left open at the end of Carmen's presentation. Ambiguities can be eliminated by convention. Ann decides that as long as she is designing the number system, she can make up the rules!

Ellen now turns to Sharon's group.

"We started like the first group," Sharon begins, "with *0, A, AA, AAA, AAAA, B.* And the flat was *C,* and the big cube was *D.* But, like Roger said, it was too cumbersome to write things that way. We were writing things like *BBAAAA* for 14, a pretty small number, and we didn't like writing all that."

Dissatisfied with the limitations of their initial system, Sharon's group constructed an alternative to address them. (See Figure 3.5.) "So we came up with the symbols to show how many *A*'s and how many *B*'s. Instead of writing *BBAAAA* we write: *BbAd.* Because *A* is now *Aa, AA* is now *Ab, AAA* is now *Ac,* and *AAAA* is now *Ad. BB* is *Bb.*

"We also decided to use *0* to hold a place. For example, in *Ca0Aa* we use *0* to hold the *B*'s place. *Ca0Aa* would be a *C,* no *B*'s, and an *A*—a flat, no rods, and a small cube. That was *Z* in the old system."

"So it can't be written as *CaAa?*" asks Ellen. "It has to be *Ca0Aa?*"

"Well," Sharon replies hesitantly, "we said it has to have the *0.* I'm not sure why. Like Ann, we made that decision."

Just as Lyanne had done in her small group, Sharon, too, is mimicking a feature of the place-value system by inserting *0* in the expression. And again like Lyanne, she has missed the essence of place value since, as Ellen's question points up, the place holder in her system is superfluous. But then Sharon goes on to explain how she came to realize that there was still something missing from her system:

"One thing that I realized this morning was: we *thought* we had place value, but I don't think so anymore because no letter moves out of its place. *A* never moves out of its place on the right, and *C* never moves from here, and *B* never moves from its place. So I think, now looking at this, I see how we've limited ourselves. I'm not sure how to change it, but it should be more flexible."

In fact, now that she is aware of this limitation, Sharon is not far from transforming her current system into one with place value. Realizing that, Deborah questions her further. "What would happen if you just erased the *A*, *B*, *C*, and *D* and left only the small letters?"

"Oh, yeah. I *could* write *b0c* and get a number. It would mean this many [two] flats because the *b* is in the third column, no rods, and this many [three] small cubes. Yeah. It's a number. Originally we would have written it as *CCAAA*."

Because *A*, *B*, *C*, and *D* are always in the same column, when Sharon sees that they are redundant and can be eliminated, she has, in effect, a base-five place-value system. However, even as she revises her conventions, she does not yet recognize this. She needs time and privacy to explore her new creation, to identify its properties, and to figure out how to operate in it.

Next, Julie describes the circuitous route by which her group arrived, in the end, at a base-five place-value system.

"We also started with this," she begins, pointing to Carmen's chart. "We struggled for a long time to make it work. We were pretty satisfied, and then Jim came along and asked us what the largest number we could write was. Well, we could do that. It was *DDDDCCCCBBBBAAAA*. And then he asked, 'What's *A* more than that?' Well, our rule was that we couldn't write more than four of a symbol in the number, and *A* more would be *DDDDD*. So then we tried doing things like *(B)D*. The parenthesis told you how many *D*'s. It was sort of like what Sharon's group did, but our notation ended up getting very confusing, so we eventually scrapped it. Then we tried to use a place-value system, with ones, tens, and hundreds columns."

As we have seen with the other groups, Julie's had also tried to mimic the base-ten place-value system, but did not see which essential features translate and which do not.

"We tried to use ones, tens, and hundreds columns, but since we were using only *0*, *A*, *B*, *C*, and *D*, it wasn't fitting in. We could only count up to four and didn't know how to get to ten for the next place. So then we tried to use place values with ones, fives, tens, twenty-fives, and hundreds columns."

Each time Julie's group discovered a limitation in their system, they designed a variation to overcome it.

"When we tried to represent *A0B*—that was one ten, no fives, and two ones—with the blocks, well, it didn't work. The number was twelve and we needed two rods and two cubes, and that didn't match what we had in our columns. So we knew we had to get rid of the tens and hundreds columns, anything that wasn't represented by the blocks. So this is what we finally got to. We have ones, fives, twenty-fives, and one hundred twenty-fives columns." (See Figure 3.5.)

Now the group has designed a system that satisfies their requirements, and they show the class how to count in it.

"*0* is the place holder. Now we count," Julie holds up an additional finger for each item, "*A, B, C, D, A0*." She pauses to make sure everyone got that *A0*. Then she continues, "*AA, AB, AC, AD, B0*." Her partner jumps up to contribute her fingers as Julie holds up both of her hands. "*BA, BB, BC, BD, C0, CA, CB, CC, CD, D0*.

"Here's how to add in this system. Take *DD* + *A*. *DD* means four rods and four cubes. You add one more cube and you've got five cubes. That means you gather them and trade them for a rod. But now you have five rods, so you change them for a flat. That's what you're left with. One flat, no rods, and no cubes. *A00. DD* + *A* = *A00*."

Julie now explains what she understands about place value.

"When Sharon said her problem was that the letters never change places, that really clicked for me. That's what place value is. The symbols have to change position. And they change value when they change position."

"One of the things that was satisfying," Julie says, "was, once we got to the place-value system we realized we could represent any quantity. And that was the frustration with every other system we tried along the way. In the other systems there were always quantities we couldn't represent."

Since the remaining groups have constructed systems that are essentially identical to one or the other of those already presented, the class goes on to a general discussion. Some time is spent talking about the experience of working in groups, of performing the task, and of hearing about other groups—how they worked and what they constructed. In their own processes, a number of the teachers can now recognize their students. In the context of the discussion, staff look for opportunities to point out how some of the Xmania systems resemble those of ancient cultures. Then participants return to their small groups.

CONTINUED EXPLORATIONS

Through the group presentations, the teachers have been exposed to four different systems all designed under similar constraints. By the end of the discussion, they agree that the place-value system is the most efficient of the four. Now the class needs time to explore these ideas in the greater privacy of their small groups, each one posing its own questions.

For those teachers who constructed systems without place value, listening to their classmates (or their instructors) explain so complex an idea is generally

Complete the following table according to the new, improved Xmanian number system.

A	B	C	D	AO
AA	AB	AC		
	BB			
DA				

Fill in the squares below as though the patterns were laid on top of an extension of the Xmanian table.

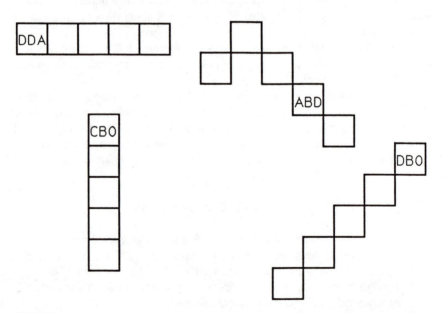

FIGURE 3.6. An activity designed to help participants examine the structure of the Xmanian place-value system

not enough to induce real understanding. These teachers will still need to engage with the symbols and blocks, this time with clues obtained from the group discussion; they must "muck about" in order to construct the concept of place value for themselves. In fact, they must grapple with a big idea analogous to the one young children work on—that within the same numeral, a particular digit, say *B*, might represent *B* units, or *B* groups of 5, or *B* groups of 25, and so on. To help them begin to count in base five, the activity sheet in Figure 3.6 is distributed, and when they are ready, they will explore addition and subtraction.

(The reader, too, is encouraged to work on the charts in Figure 3.6. If you had trouble following the descriptions of the various systems, then, after working on the charts, you might want to reread Julie's explanations.)

For those who have already created a place-value system, there is much more to explore. For example, James, Louise, and Patty decide to try calculating *AA* × *AA* and arrange the blocks to represent the problem. Then they trade to get an answer of *ABA* (as shown in Figure 3.7). Curious as to what patterns they will find, they next arrange the blocks to represent *AA* × *BB* and then trade to get *BDB* (as shown in Figure 3.8).

James becomes very excited: "Look, there is a pattern happening! The middle number is double the end numbers . . . hmmm . . . but that's only going to work until we have to trade up more. *AA* × *CC* will require a further trade, but the pattern does hold if you don't trade completely." He lays out *AA* × *CC* to demonstrate that the number of sticks is double the number of flats and the number of cubes, which are equal (see Figure 3.9).

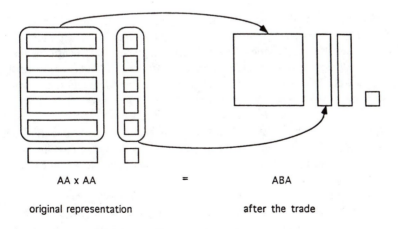

AA x AA = ABA

original representation   after the trade

FIGURE 3.7. Manipulation of blocks to represent AA x AA

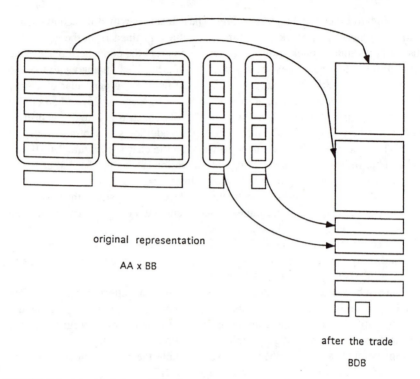

original representation

AA x BB

after the trade

BDB

FIGURE 3.8. Manipulation of blocks to represent AA x BB

Louise comments that she sees a different pattern emerging. She separates one of the CC groups (see Figure 3.10) and explains that it is a representation of $A \times CC$. She then pushes the remaining pieces together, and, measuring them with a stick, demonstrates that the remainder is $A0 \times CC$. "What we're really doing," she declares, "is multiplying two parts of the problem and then adding them together. In this case we are adding $(A \times CC) + (A0 \times CC)$, which equals $CC + CC0$. It's the distributive property in action!"

"Oh, right!" exclaims Patty. "You know, this is embarrassing to admit, but for years I've taught children that when you multiply 10 times anything, you just add 0 to the number as a place holder. Until this moment, I don't think I fully understood why."

Ellen, having just joined the group, hears Patty's admission and asks her to explain what she now understands. "Well," she begins, "when you multiply an amount by any amount of sticks, the units will always be an even amount for

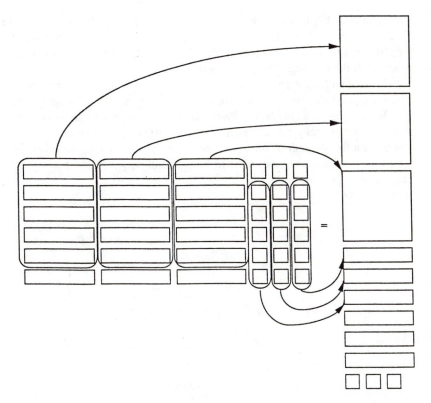

FIGURE 3.9. AA x CC with incomplete trade producing 3 flats, 6 sticks, and 3 cubes

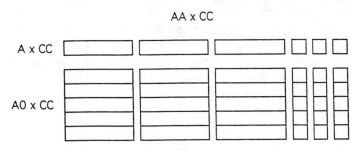

FIGURE 3.10. Demonstrating distributivity: AA x CC = (A x CC) + (AO x CC)

trading into sticks. In this case the sticks are worth 5 units, and since 5 × 3 is the same as 3 × 5, the units are equal to 3 sticks. Because of the commutative property, it will always be an even trade with no units left over." She demonstrates with the examples shown in Figure 3.11.

Pleased with their progress so far, the group decides to use Louise's method of sorting problems into parts and tries $AB \times ACB$ (illustrated in Figure 3.12).

Working in Xmania, James, Patty, and Louise are able to discover patterns and explain why they work. Parallel patterns in base ten have been taken for granted, on the strength of someone else's authority, for many years now. Their discoveries in base five, translated to base ten, provide familiar rules of computation with new meaning.

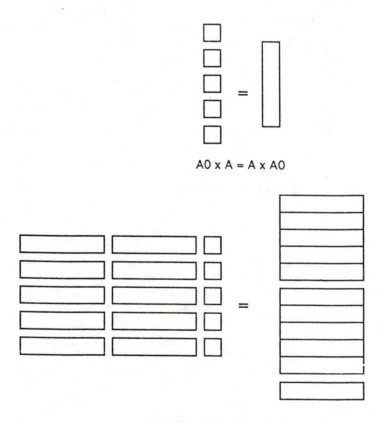

AO x A = A x AO

AO x BA = BA x AO = BAO

FIGURE 3.11. Demonstrating commutativity

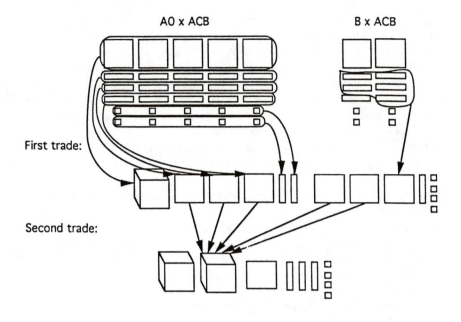

AB x ACB = BACD

FIGURE 3.12. AB x ACB with base-ten blocks

REFLECTIONS ON LEARNING

The Xmania activity offers teachers the opportunity to endow the symbols and procedures of our number system with meaning. Since the activity involves a base-five place-value system, they cannot rely on remembered procedures that may have no conceptual depth for them. However, once the logic of that system has been explored, many teachers can translate their insights back into the system they teach.

We have explored Xmania with many groups, and it is interesting to note how consistently their activity recapitulates human development in two ways. First, many teachers hit upon the same kinds of ideas, and make the same kinds of errors, as children who are learning the base-ten place-value system. Second, the systems that the teachers initially design usually resemble those of ancient civilizations.

The Xmania activity is meant to be accessible to everyone, requiring only superficial understanding of the number system commonly used. As the activity continues, those teachers who need to spend more time figuring out just how the symbols of a place-value system correspond to the quantities they represent, who need to construct meaning for place value, can do so. Those who feel secure in their understanding of the place-value system can extend their investigations limitlessly.

Many groups, such as the one comprising James, Louise, and Patty, pursue questions that they pose for themselves. At times, program staff suggest others: How can you tell if an Xmanian number is even or odd? What are the rules of divisibility for *C, D, A0,* and *AA*? What are fractions in Xmania, and what would correspond to decimals? For which fractions are the equivalents "decimals" (or should we call them "pentimals"?) terminating? What are their analogs in base ten?

While each group collectively works on the same problem, group members are not necessarily constructing the same understandings. For example, James's group illustrates how, when the same mathematical situation is viewed through the lenses of their different interests, its members acquire different mathematical insights: while James is focused on the patterns created when multiplying by *AA*, Patty sees new meaning for the results of multiplying by ten, and Louise comes to understand the relationship between distributivity and multiplying large numbers. Other participants in the exercise also comment either orally or in writing on what they learned.

Sharon compares her new knowledge of place value to what she had previously been satisfied to consider mathematical understanding.

I've taught place value over and over and over again and I've told the kids, "We only have ten numerals and the way the number system works is, the place tells you the value of the number." I've said it a hundred times and here I went to design a system and I couldn't use the methods that I tell people over and over again. So I do feel like it was a major thing that I learned. . . . It was worth the frustration to get what I think is a lasting understanding of place value.

Betsy, having discovered the power of her own thinking, continues to pose her own questions as she further explores Xmania.

When we had counted, created an awesome place-value system, then learned all 4 operations and done our homework of preparing the charts, then I wanted to figure out fractions. . . . Then on the way to dinner, Dennis and I worked on decimals and figured them out. Then after dinner Pat and I figured out time. It was *awesome*! I've never had so much fun, actually figuring all this out by myself! I really got it! I did indeed! I even helped Ann get past her stuck point with it! So I knew I knew it well! But that wasn't enough. I wanted to know more about it, so I kept thinking of more ways to use this new language and when I got to my room, I had to call my 2 best friends in Amherst and describe it to them. My cup runneth over!

In contrast to the exhilaration that Betsy feels, Ana finds Xmania much more of a struggle.

Our group presented the first system. We had some difficulty explaining. Once we presented, the group knew there was some trouble with the system. Working on this project I had many different kinds of emotions, but at [this time] I felt frustrated enough that I just wanted to get it over with. [But later,] once we came back to the group to work on some Xmania problems [in the base-five place-value system], Jim came to help us with the exercise. I was glad he asked us some questions. Then I was able to figure it out. . . . I really enjoyed the [last] whole hour.

Donald also has the group presentations in mind as he considers what he has learned about number systems in general.

Having us look at similarities and differences between Carmen's group and Ann's group really helped me define each group's system's essential features. Thinking about questions such as, What are the conventions? Does order matter? also helped me focus on essential qualities of any number system, and I'm trying to formulate connections to how kindergartners develop their sense of number.

Meanwhile, Linda is intrigued by the relationship between her own route to the creation of place value and the historical development: "Starting from scratch was like a crash course in the history of the development of our number system!!! *Wow!!*"

And Ginny learns to sympathize with her students' difficulties as she recalls her own errors.

Working . . . on a two-by-one-digit multiplication problem, I can't believe I forgot to regroup from the units to the *A0*'s. Amazing! I should now have more sympathy for my third graders when they do similar actions with tens.

As this entry from Judy's journal shows, some of the teachers are acquiring the rudiments of an understanding of constructivism.

Working in the other base was such an eye-opener for me. I knew manipulatives are valuable—now I really *know*. I couldn't work with the symbols only. I couldn't see the pattern or handle a simple subtraction operation. I was a kind of first-grade student, I think. I also had to [develop] my understanding as I went along with the manipulatives. I just realized as I write this that that was what I was doing—making sense of it.

Even teachers who began the program with a theoretical commitment to constructivism find that the Xmania experience gives new meaning to their beliefs. Julie writes:

I have at various times in the past seriously studied Piaget . . . yet I never *really* knew what they meant by "construct knowledge from within." I'm getting it now because [I'm aware that I] have . . . construct[ed] knowledge for myself.

What is notable about all of these reflections on a mathematics activity is how present the learner is. The mathematics these teachers describe is intensely close to them. It is exhilarating, frustrating, surprising, puzzling, satisfying, intriguing, and "awesome." They build it, ponder it, model, discover, and make meaning. It engages their creativity and is esthetically pleasing. To them, mathematics—at least in the form of the Xmania exercise—is not a cold, distant set of facts, rules, and procedures handed down from on high. Mathematics, they are beginning to see, is *their* construction, too, as individuals, as participants in the community that is the mathematics class, and as members of the human race.

4
Becoming a Mathematical Thinker: Linda Sarage

Central to the preparation for teaching mathematics is the development of a deep understanding of the mathematics of the school curriculum and how it fits within the discipline of mathematics. Too often, it is taken for granted that teachers' knowledge of the content of school mathematics is in place by the time they complete their own K–12 learning experiences. Teachers need opportunities to revisit school mathematics topics in ways that will allow them to develop deeper understandings of the subtle ideas and relationships that are involved between and among concepts.

NCTM, *Professional Standards for Teaching Mathematics*

One considerable obstacle to widespread adoption of the new paradigm of mathematics instruction is that most teachers do not understand school mathematics well enough to promote significant mathematical exploration in their classrooms. Themselves products of traditional mathematics education, these teachers doubt their own ability to think mathematically, and view mathematics as a mystifying sequence of facts, definitions, and rule-governed procedures.

The idea that SummerMath for Teachers should offer a mathematics course specifically designed for practicing teachers originated with program participants themselves. Among the local teachers who had been to an institute, many were becoming increasingly aware that their mathematical knowledge was too superficial to allow them to teach as they now wished to. They nonetheless rejected those courses already available at area colleges and universities. "We need a math course taught the way you're teaching us to teach," many commented.

This chapter follows Linda Sarage, a third-grade teacher from a rural school district, as she progresses through the mathematics course that resulted from those requests. The development of Linda's mathematical thinking is examined, with particular emphasis on how the cognitive and affective aspects of this process are inextricably bound to one another.

COURSE STRUCTURE

The course—designed and taught that first spring by Deborah Schifter— has three major components: mathematical explorations, reading assignments, and journal keeping. Each session begins by offering participants the opportunity to bring up any thoughts or questions about the previous class, the mathematics homework, or the reading assignments. Opening discussion usually lasts a half hour to an hour, leaving the balance of class time for mathematics explorations whose format—as illustrated in Chapters 2 and 3—usually involves working on an activity in small groups and then sharing discoveries and questions with the whole class. Activities are designed at once to stimulate the construction of powerful—"big"—mathematical ideas and to model alternative instructional approaches. Homework includes further problems related to the mathematical explorations, reading assignments of an article or two, and journal keeping.

The mathematics chosen for exploration is selected from the elementary- and middle-school curriculum: whole-number operations, integers, fractions, decimals, exponents, area and perimeter, and properties of geometric figures. But, on occasion, class discussion extends to such "advanced" concepts as limits and non-Euclidean geometries.

Reading assignments, distributed throughout the course on a weekly basis, address such topics as classroom practice informed by constructivist principles (Davis, 1966; Finkel & Monk, 1983; Hammerman & Davidson, 1992; Lester, 1989; Narode, 1988; Simon, 1986), affective aspects of mathematics learning (Goldin, 1988), cognitive and metacognitive processes (Crowley, 1987; Schoenfeld, 1989), and mathematical misconceptions (Clement, Narode, & Rosnick, 1981), in addition to instructional approaches to selected mathematical ideas (Lampert, 1989; Ross, 1989). The readings are chosen to help teachers interpret their own experiences in the course and to suggest how these can be translated into their own classrooms.

Journals are used as a means of reflection as well as a vehicle for dialogue. Teachers write about what they are learning and what they find interesting, the ideas and concepts with which they are currently struggling, events in their own classes, and their personal reactions to course activities. The instructor collects journals once a month and responds to them in writing before returning them the next week.

The course is open to teachers who are just entering the program as well as to teachers who have already attended an institute.

LINDA'S BACKGROUND

Linda enrolled in the course the first year it was offered. She had not previously attended an institute, but among Linda's colleagues were several SummerMath for Teachers participants who had conducted in-service workshops in her school. In her application essay, Linda wrote:

I have begun to use some of the methods these teachers have shared informally with me . . . and I'm convinced that the developmental basis of constructivist math education is fundamental to effective math instruction and learning. I have been dissatisfied with my past efforts at teaching math. I've known that mathematics should not be a dull, dry discipline— that math encompasses far more than simply manipulating numbers. But since my own math conceptual framework is very weak, I need to rely on prepared curriculum texts, workbooks, and practice sheets. . . . I welcome the opportunity this course offers to explore math concepts in depth. I value, too, the chance to learn more about the methods of constructivist education.

At the age of 36, Linda had been teaching for three years. After her children had started school, Linda herself entered college in a program for nontraditional students. There she received teacher certification and was now teaching third grade.

Linda's description of her history as a mathematics student is a familiar one. Many SummerMath for Teachers participants, especially the elementary-school teachers, tell similar tales:

I was always very afraid of [math]. I mean, I didn't do well in math in high school. I barely remember; it's like I blocked it out. Then when I went back to college I took one math course called Contemporary Math. It was just supposed to be practical—just to balance the checkbook. And I excelled in that math class. I got an A+ for a final grade. But the anxiety that came with that math was incredible. . . . And then I ended up not having to take any more math in college, and I was really relieved.

Now, six years later, here she was, enrolled in Fundamental Concepts for Mathematics Instruction.

GIVING MEANING TO SYMBOLS

Where traditional paradigms of mathematics instruction prevail, students from first grade on are generally taught a variety of formal expressions and operations whose connection to their informal mathematical knowledge is never established. When the primary focus of concern is that students should be working in "the abstract"—"doing real math"—teachers, parents, and students are likely to remain unaware that these expressions and operations derive their meaning from conceptual structures ultimately rooted in, *abstracted from*, experience.

By contrast, for students in, say, Ginny Brown's classes, the words and symbols that designate division (" ÷ ," "divided by," "quotient," "remainder") represent conceptual structures that they themselves have derived from the experience of sharing (distributing objects into equal-sized groups and deciding what to do with what is left over). Because these third graders move fluently among abstract notations, concrete embodiments, and images of concrete embodiments, they are able to extend their mathematical understanding in the face of new problems.

Most of the teachers who enter the program are themselves products of traditional mathematics education; the mathematical formalisms they have been taught are largely devoid of meaning for them (Ball, 1988; Leinhardt & Smith, 1985; McDiarmid, Ball, & Anderson, 1989). Thus, while they are familiar with the algorithms for the basic mathematical operations, they need to begin to make meaning for them in order to develop the same kind of facility that Ginny's students possess. To this end, the first lessons in the spring course are organized around explorations of the properties of the number system. The challenge is to find concrete embodiments demonstrating general principles. For example, the activity sheet reproduced in Figure 4.1 is distributed to the class following an initial discussion. Teachers work on it in groups of three.

Where these activities extend to such concepts as negative number or the meanings of "inverse" and "reciprocal," major emphasis is given to representing

Figure 4.1 Activity Sheet to Explore Properties of the Number System

---

WHAT'S YOUR ORDER?

With some operations,* order doesn't make a difference. For example:
    $3 + 4 = 4 + 3 = 7$
With other operations you get a different answer when you change the order of the numbers: $3 - 4$ does not give the same answer as $4 - 3$.
The task today is to investigate when order makes a difference and when it doesn't.
(1) For what operations does the order of the numbers not make a difference? Use manipulatives, diagrams, or real-world situations (story problems) to demonstrate why.
(2) For what operations does the order of the numbers make a difference?
    (a) Are there special cases where it doesn't make a difference? What are they?
    (b) When it does make a difference, what patterns do you see in the different answers?
    (c) Are the patterns just special cases, or do they hold for all pairs of numbers? Demonstrate using manipulatives, diagrams, or real-world situations.

*By operations, we mean addition, subtraction, multiplication, and division.

---

patterns with diagrams, manipulatives, or story problems. Such exercises lead teachers to a growing awareness of the need to make meaning for familiar symbols and operations.

As Linda grappled with these issues, she wrote in her journal: "I don't have any idea what $8 \div 12$ means in the real world. I know the answer is ⅔. Does that count?" Even as she repeated phrases uttered by her classmates, she was aware that they were contentless for her. She had not constructed their meaning for herself:

> 8 whole candy bars divided by 12 pieces means each piece is ⅔ of a whole bar. But I know I'm just substituting words and numbers. I need to spend more time looking at it. Maybe translating $\div$ as "divided by" is the problem. 8 whole candy bars sorted by 12 pieces—no, no.

Linda listened to her classmates, trying to figure out what was helping them to greater understanding, but their explanations and diagrams were just as incomprehensible as the original questions. "I still am baffled about how they knew to solve the $8 \div 12$, $12 \div 8$ problem. Apparently, the picture was the key for those who understood the difference."

By this time, Linda was aware of the possibility that she could make sense for all these empty words and symbols, but she was still unable to connect formalism with meaning. "If I keep playing," she wrote,

> the interrelationships between the operations will become more and more tangible for me. I can say—subtraction is the reverse of addition—but that's a limited way of looking at that operation. Besides, I guess I don't even know what "is the reverse of" means—*and* what the implications may be in various situations.

The lesson sequence continued with similar explorations of the associative and distributive properties, again emphasizing concrete representations. Having completed a homework assignment on the distributive property, the class reported that it held in the following cases:

a. $(4 + 2) \times 12$ and $(4 \times 12) + (2 \times 12)$
b. $(4 + 2) \div 12$ and $(4 \div 12) + (2 \div 12)$
c. $12 \times (4 + 2)$ and $(12 \times 4) + (12 \times 2)$

but not in:

d. $12 \div (4 + 2)$ and $(12 \div 4) + (12 \div 2)$.

The class's feeling about this finding can best be described as mystification. "As I have played with this week's distributive property of multiplication (and non-distributivity of division) the *why* question haunts me," Linda later wrote. Why *should* the pattern hold in cases *a–c*, but not in *d*? The arrays of symbols had no meaning for her.

As the class mulled over the assignment, one teacher, thinking about previous lessons, suggested that they make up word problems for each expression. For *c*, the class suggested, "There were 4 boys and 2 girls, and each child had 12 candy bars. How many candy bars were there altogether?" They were satisfied that both expressions fit the word problem. And for *d* they suggested: "There were 4 boys and 2 girls who had 12 candy bars to share among themselves. How many did each child get?" Now they saw that that fit the expression on the left in *d*. For the expression on the right they began by analogy: "There were 12 candy bars to share among 4 boys and another 12 to share among 2 girls . . ." Suddenly, there were several gasps and "oh's" in the room. "It's a different situation!" The concrete context gave meaning to the symbols, meaning that offered grounding, access, and a sense of ownership over the ideas.

Linda described her experience: "Seeing the division example as a word problem was [mind-]boggling. Suddenly the 'Why won't it work?' appeared so clear." She was learning to identify what understanding feels like when mathematical notation suddenly connects to something already solidly known.

After this breakthrough, Linda began to develop fluency in examining familiar mathematical algorithms in the context of word problems, manipulatives, and diagrams. Her progress could be observed when, several weeks later in the middle of a unit on fractions, Deborah gave the class a division problem: *Four sheet cakes were divided into portions of three-fifths of a cake. How many portions were there?*

                                   *   *   *

"I'll show my diagram," Vivian volunteers, going to the board to demonstrate her solution to the rest of the class. She draws the diagram shown in Figure 4.2. "See, here are the four cakes. And you can see that you get six portions out of them."

"What about the left-overs?"Clare asks.

"Okay, so it's 6 and ⅖," Vivian responds after a glance at her diagram.

Wait a minute," Jonathan interjects. "When I do the calculation I get 6 and ⅔."

"But look at the picture," says Vivian. "You can see that it's ⅖ left over."

Seeing the significance of the "contradiction" Vivian and Jonathan have found, Deborah interrupts their debate to make sure the rest of the class sees it, too. "I'd like everyone to find a partner to discuss what's going on here."

The class is puzzling over what to do with the two pieces of cake that are

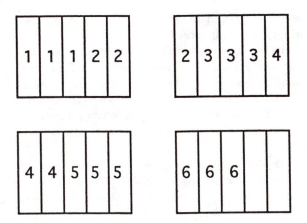

FIGURE 4.2. Four cakes divided into portions of 3/5

left after six people take their portions. The confusion comes about because those two together comprise ⅖ of a cake, but ⅔ of a portion.

"I get it!" Carole declares as the class comes together again. "The problem asks about portions. You can say that there are 6 portions with ⅖ of a cake left over. Or you can say that there are 6 and ⅔ portions."

"Oh, yeah." Then the class is quiet a moment, letting the idea sink in.

"Look at something else in Vivian's diagram," Sandy urges. "She started with 4 cakes. Then she cut each cake into 5 pieces. So she has 4 × 5 = 20 pieces. Then she grouped those pieces by 3's since 3 pieces make up a portion. So she got 20 ÷ 3 = 6⅔ portions. She multiplied by the denominator and divided by the numerator. Like in flip and multiply!"

"I see something else in that diagram," Eleanor adds. "You've got ⅗ of a cake equal to a portion. But you can see that each cake is one portion plus another ⅔ of a portion. That's each cake is 5/3 of a portion. So when you want to find out how many portions there are in 4 cakes, you can divide by the size of each portion, or you can multiply by the number of portions per cake. Amazing! You can answer the question by 4 ÷ ⅗ or by 4 × 5/3. I never knew that could make so much sense!"

Both Sandy and Eleanor are describing the meaning they now find in the hitherto mysterious rule for dividing fractions.

*   *   *

At the time, Linda was unable to see what her classmates saw and went home to explore their insights further. She began with the bits of information she had picked up from the class discussion.

"I can see 4 × 5/3, with 5/3 = each cake's portion. I can see 4 ÷ ⅗, with

$\frac{3}{5}$ = each person's portion. I do not know how this problem can be generalized to explain the flip-and-multiply rule!"

At this point Linda was beginning to grapple with a big idea of fractions—that the quantity represented by a fraction depends on the reference whole, and that, depending on the whole, a single quantity might be represented by different fractions. She said that "each cake's portion" is $\frac{5}{3}$ and "each person's portion" is $\frac{3}{5}$. But she did not yet see that the whole to which $\frac{5}{3}$ refers is a single portion while the whole to which $\frac{3}{5}$ refers is an entire cake.

> Wait a minute—in order to know how many $\frac{3}{5}$ portions there were in 4 cakes, I ended up multiplying the number of cakes by the number of portions each cake had! Which was $4 \times \frac{5}{3}$. Does this make sense? If $\frac{3}{5}$ is a portion, then $\frac{1}{5}$ of a cake is $\frac{1}{3}$ of a portion.
>
> | *Cake* | | *Portions* |
> |---|---|---|
> | $\frac{2}{5}$ | = | $\frac{2}{3}$ |
> | $\frac{3}{5}$ | = | $\frac{3}{3}$ |
> | $\frac{4}{5}$ | = | $\frac{4}{3}$ |
> | $\frac{5}{5}$ | = | $\frac{5}{3}$ |
>
> Each cake has $\frac{5}{3}$ portions.

Now Linda was beginning to understand that the whole the fractions in the first column refer to is one cake and that the whole the fractions in the second column refer to is a single portion. She found the results quite shocking: "My God—if anyone had ever told me that $\frac{5}{5}$ could be equal to $\frac{5}{3}$, I'd tell them they were crazy. What is happening here?"

In order to consolidate her understanding, Linda made up some problems similar in form but with different numbers and then followed the same logic to see if the result would be consistent with her solution to the original problem. She began with a diagrammatic strategy, her most "trusted" method for dividing fractions, and compared it to her "flip-and-multiply" logic:

> 3 cakes divided into $\frac{3}{4}$ portions. How many portions? [See Linda's diagram in Figure 4.3.]
> $3 \div \frac{3}{4} = 3 \times \frac{4}{3} = 4$ portions.
> If $\frac{3}{4}$ is a portion, then $\frac{1}{4}$ of a cake is $\frac{1}{3}$ of a portion.
> $\frac{4}{4}$ of a cake has $\frac{4}{3}$ portions.

And again:

> 6 cakes, each portion $\frac{2}{5}$. How many portions? [See Linda's diagram in Figure 4.4.]

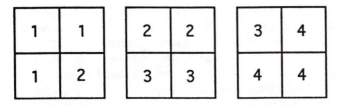

FIGURE 4.3. Three cakes divided into portions of 3/4

1		3	6	8	11	13
1		4	6	9	11	14
2		4	7	9	12	14
2		5	7	10	12	15
3		5	8	10	13	15

FIGURE 4.4. Six cakes divided into portions of 2/5

If ⅖ is a portion, then ⅕ is ½ portion.
⅖ (whole) is ⁵⁄₂ portions for each cake.
$6 \div \frac{2}{5} = 6 \times \frac{5}{2} = 15$ portions.
Generalized Statement—The inverse of the fraction indicates the number
of portions per whole!! I did it??

When one reads the record of Linda's thinking process in her journal, questions arise concerning just what she understood. In her "generalized statement,"
what was it that "the fraction" represented? And did she really understand the
nature of the divisor in a division-of-fractions problem?

What is clear, however, is that Linda had learned to devise specific examples in order to examine a given mathematical issue. She imagined concrete
situations—concrete in the sense that dividing cakes and sorting them into portions of equal size is experientially grounded for her—and used diagrams (which
earlier in the course had had no meaning for her) to display information and
expose relationships. Once enough examples had been investigated and the
underlying pattern revealed, she worked to convince herself of the generalization
that could be made based on the mathematical structure that the specific examples shared. In this way she was able to make sense of a rule that until now
she had been forced to accept on authority (Burton, 1984; Mason, Burton, &
Stacey, 1982).

In her journal the following week, Linda reflected on the fractions unit just
completed:

I have probably worked more with fractions in the last four weeks than I have in the four years preceding this! Do I have a better understanding of fractions? Definitely. What do I understand better?

Fractions have a real meaning. [For example,] I can now see ⅓ of ⅙. This seeing gives the ¹⁄₁₈ a different form. But can I put this into a word problem? Sure! I cut a cake into 6 pieces. Each piece is ⅙ of the cake. I'm going to put purple frosting on ⅓ of a piece. ¹⁄₁₈ of the whole cake will have purple frosting.

Now, like the children in Ginny Brown's class, Linda could fluently move between abstract notations of mathematical concepts and physical models of those same concepts. This fluency allowed her to operate on both, each representation informing the other. (For further discussion of adults' understanding of fractions, see, for example, Ball, 1988; Harel, Behr, Post, & Lesh, 1989; Simon, 1990.)

COPING WITH CONFUSION

Before continuing to track Linda's mathematical development, it may be helpful to retrace her experience of the course so far in order to illuminate her affective process. But first imagine a traditional classroom setting where good teaching is understood to be clear explanation. The teacher has just finished explaining division of fractions to her seventh-grade class, and Jan, one of her students, looks at the board (see Figure 4.5) in despair.

Jan doesn't know what to make of all the symbols, not even the first step. She looks to her right and left and, seeing her classmates busy with the page of division-of-fractions exercises the teacher has distributed, assumes that they understand what they are doing. Only she is confused. And she knows that if she asks a question, she'll just get the same confusing response. The teacher is very patient and more than willing to repeat herself, for she relishes the beauty of her own explanation. But hearing the same words one more time will not address Jan's difficulty.

In this class, where evidence of learning is getting right answers on a page of calculations, "making meaning"—in the sense described previously—appar-

$$6 \div 3/5 = \frac{6}{3/5} = \frac{6}{3/5} \times \frac{5/3}{5/3} = \frac{6 \times 5/3}{1} = 6 \times 5/3$$

FIGURE 4.5. A demonstration of the equivalence between dividing by a fraction and multiplying by its inverse

ently has no role. The essence of learning is following directions and remembering them. There is no room for confusion.

For contrast, let us once again return to Ginny Brown's classroom. There, confusion was the prelude to learning. The children were puzzled by a division problem in which the distributed blocks did not yield groups of equal size. Faced with this dilemma, they were forced to extend their conception of division and came up with two possible solutions to the problem. In Piagetian terms, their experience of disequilibrium necessitated a transformation of their current understanding of division in order to accommodate that conception to the novel situation. The result was the construction of a more inclusive concept—division with remainder.

Similarly, the class of mathematics teachers had been confused by conflicting answers to the sheetcake problem. The two solutions, 6⅖ and 6⅔, had been obtained through different solution processes. This prompted teachers to reexamine the problem until they realized that the same quantity could be represented by different fractions, depending on what they considered to be the whole to which the fraction referred. This insight led, in turn, to new interpretations of the familiar flip-and-multiply rule.

The confusion, or disequilibrium, that all too many students like Jan experience could be turned to stimulate the construction of meaning for the notation that covers the board each day. However, their classes are not structured to lead them to resolution of their confusion, and their well-intentioned teachers do not know how to help. Thus, Jan's confusion indicates both to her and to her teacher that she has failed to learn how to perform the operation that she has been shown. Jan has learned that she cannot learn mathematics.

From Linda's description of her own history as a mathematics student, we may assume that it was similar to Jan's. In a journal excerpt written early in the semester, it is apparent that Linda had learned that same lesson about the meaning of confusion. "Two weeks into class," she wrote, "and already I'm getting confused. Actually I guess I'm more nervous about getting confused than troubled about my current confusion level."

As Linda's instructor, Deborah valued the mathematical questions Linda was posing at the time. She felt that Linda was looking for real-world meaning for "8 ÷ 12," and thought it important that Linda wondered what it "really" means to say that addition and subtraction are inverse operations. Furthermore, Deborah knew that it was also important that Linda find her *own* answers to these questions. But for Linda, the class period "seemed to drag. I could sense an 'I don't care' attitude within."

During the next class, Linda's confusion was compounded by her sense of exposure and isolation: "I felt really stupid. . . .I was surprised to find myself fighting back tears. And I wanted to ask more questions, but I felt that I said I didn't understand enough already."

It was not only mathematics content that Linda and her classmates needed to learn, but also something about the learning process itself, including its affective aspects. To this end, Deborah had earlier distributed an article about affective states in problem-solving processes. In it, the author, Gerald Goldin (1988), suggests that in order for a problem to be felt as a problem, the erstwhile problem solver must experience puzzlement and, quite possibly, as solution strategies are pursued, bewilderment. When a solution is finally hit upon, one experiences satisfaction, but if initial attempts are fruitless, frustration results. However, "frustration need not be a negative problem-solving outcome" (p. 5), but may instead become a spur, goading one to try other avenues of approach. If these, too, are unsuccessful, the result can be anxiety or even despair, and a more urgent problem—how to get out of an unbearable situation—may supplant the original. When this happens, the problem solver may drop out, make a guess, or just mimic someone else's solution without understanding it.

Linda read the article with relief at recognizing herself:

> The article makes *so much sense* and I could see myself throughout. Unfortunately the negative effect schema is how I see myself. . . . Even the avoidance—see, two days before class and I'm literally forcing myself to get started on homework. If she would just tell me the right answer, I'll accept it on authoritative grounds. But there's hope! Thanks for this article. I'm actually laughing.

Linda did not give up. Throughout those first weeks, she sat in class listening, trying to sort out what had helped her classmates to understand. "Some students interchanged ideas on a level of understanding that I haven't (yet?) achieved. I still am baffled about how they knew to solve the 8 ÷ 12, 12 ÷ 8 problem. Apparently, the picture was the key for those who understood the difference."

Able now to identify confusion and frustration as stages on the way to something positive, she concluded: "So, my plan is to [keep my] hand in—try to keep playing even though I feel like I'm just pushing food around on my dinner plate—and hope that something will click."

Linda's perseverance was rewarded in the very next class period, which began with the discussion, described above, of the distributive property. That discussion Linda had called "[mind-]boggling. Suddenly the 'Why won't it work?' appeared so clear." Having felt the click of understanding, she had become excited by what she was learning and left her confusion and frustration behind.

The class moved on, in a subsequent lesson, to a series of activities based on the Xmanian number systems (described in Chapter 3), which Linda had already encountered in a workshop held in her school some months earlier. Her experience was immediately rewarding.

> On to Xmania—It was comforting to not feel like a beginner with the system. I was encouraged to be able to hold my own with others. . . . Actually what I felt was relief. The adding, subtracting, multiplying was challenging—but it was fun. Like a puzzle.
> And when I got home after class I didn't want to stop—on to division and wrestling with even numbers in Xmania.

Linda was no longer so distressed when she met up against areas of confusion. The sense of camaraderie she felt with her group actually helped her to delight in discovering the errors she was making.

> I could keep identifying with a young child trying to make sense out of our base-ten system. When we multiplied we carried the wrong letter!! And I forgot to add when I would make a trade!!! I have seen lots of students make that error, but never quite understood how come they'd do it. It's quite easy to do when trading up and down! (Forget, that is.) I also discovered that more practice with the trading helped to solidify the steps. I wanted more examples to work with.

Yet, reflecting on the lesson sequence, Linda wondered about the way the activity was structured. Couldn't it have been designed to avoid the frustration of developing inefficient number systems?

> The point about Xmania that I don't understand is that we're encouraged to find any number system that works. The point is made that no system is the correct system. But the first time I played with Xmania, my partner and I began building an add-on system, like Roman numerals. It quickly became cumbersome, although it could have been possible to develop it to the point of adding, subtracting, etc.
> When we scrapped that system and hit on a base-five, using 0 to hold place value, it felt like we discovered the "secret" of Xmania—that we were on the right track. And now I'm convinced that this base-five system is an efficient system. My goal is now to learn to operate in it. I feel no need to "explore" other possible systems. Doesn't this mean that there is a "best" solution?? And that the frustration of developing a system ends when the base-five system is realized???
> And in the end—aren't there going to be certain rules, systems, operations, we want our students to discover? And if someone had said—

These are the basic principles of Xmania; now try adding, subtracting, etc.—it still would have been fun.

On further reflection, however, Linda was able to see the value of "mucking about." The disequilibrium engendered by the lesson design resulted in learning that would otherwise not have occurred. "But wait!! Starting from scratch was like a crash course in the history of the development of our number system!!! *Wow!!*"

Continuing on to the fractions unit, Linda frequently remarked on how amazed she was to find that she could solve problems that a short while ago she would never have attempted. Of course there were still areas of weakness, yet they no longer caused the despair she had felt weeks earlier:

There's still much more to realize about dividing fractions. I know that I need to work more to solidify what I have just "discovered" before moving on to the gray areas.

Relieved of her fear of being thought wrong or incapable, Linda, now free to reflect on her own process, could consider how learning takes place.

I think my thoughts are so sketchy because the fractional thinking hasn't settled into any cognitive slots yet. They're still in the making. I know it's not enough to say I have a better "sense" of fractions. A framework is developing.

I think this is the big news for me. I don't have to get it all right now. I'm not only learning math concepts, but I'm becoming more aware of myself as a learner. Was this planned????

And this new understanding of her own learning process in turn empowered her to take new risks, to explore areas that had previously appeared inaccessible.

I guess even on this simplistic level I find myself—dare I say it?—observing, examining, thinking along mathematical lines. I used to quickly shut down if any notion appeared to be connected to "complicated" math ideas. I know what it is—confidence. Some of the mystery is lifting.

OTHER FORMS OF MEANING MAKING

As Linda and her classmates learned to make meaning for familiar symbols, they came to see that mathematics is ultimately ordered by familiar logical principles. The rules governing the basic mathematical operations are not arbitrary, need not be simply accepted, but could be demonstrated by their own powers of reasoning. And the teachers could make these discoveries themselves, communicate them, debate them, and agree on their validity.

Yet the development of mathematical systems is not, in itself, completely determined by logic. The particular designations "7," " + ," or " = " are conventional, of course, and so is the choice of 10 as the base of our number system, rather than, say, 8 or 12. But while some conventions seem totally arbitrary, others have powerful systemic ramifications. Deborah frequently looked for opportunities in class discussion to point out the complex interplay of choice, agreement, and theoretical coherence.

For example, the activity sheets for the exploration of exponents were designed so that as students worked from the definition of "whole-number exponent," they derived the product, quotient, and power rules. Once these rules had been established, students were challenged to give meaning to such expressions as 2^0 and 3^{-2}. From Linda's journal:

> The exponent work was interesting. What's even more fascinating is the notion that mathematicians have these "agreed-on" rules. I've tried to imagine any other discipline that requires a similar function. The theories of the social sciences are different. One either accepts a theory, develops it further, or rejects a theory and maybe develops another.
> So $2^5/2^5 = 2^0 = 1$ because we need to fit the "subtract exponents" rule. But what I sense is that 2^0 is a symbol—and that the agreed-on rules are for language's sake. No, it's more than that. For language's sake, scientists agree on *xyz* as the name of a newly discovered element. There's no need to fit the agreement into an existing schema.

As Linda reflected further on the dialectic between logical determination and convention, she concluded:

> So 2^4 means 2 multiplied by itself 4 times, $2 \times 2 \times 2 \times 2$. That's agreed on. Therefore, 2^0 means 2 multiplied by itself 0 times. No No No. This time the zero represents prior manipulations of the exponents. $2^x/2^x = 2^{x-x}$. So we're changing the meaning of exponent to make a rule work. But then we say 2^0 has no meaning anyway—so we'll give it a meaning to fit our rule.
> I want to come back to this someday. I think it's very convenient and very logical and very clever. I need to observe if there is a related process in any other discipline. I can't think of any right now, but I haven't been thinking along this vein before. Not ever.

With her new-found confidence, Linda could take on the question of what the expression 2^0 might mean—a question that previously would have been inaccessible—with the expectation that she could make some sense of it. More-over, by this time, finding meaning did not necessarily require physical model-ing. She was now able to work with the formal apparatus of multiplication, division, and exponentiation to analyze the convention $2^0 = 1$. And as she took her reflection to the next level, by analyzing how meaning was given to 2^0, she pondered what this kind of "meaning making" implied about mathematics in relation to other disciplines.

REVISITING THE FEAR

Linda's growing confidence did not mean that she had forever overcome her fear of mathematics. But her insight into the learning process ensured that when she again encountered that fear, it no longer had its old paralyzing power. Rather, she could recognize it as just a stage in the process.

Near the end of the semester, as the class began a unit in geometry, Linda's old anxieties resurfaced: "So we are beginning geometry," she wrote. "I haven't worked with geometry since 10th grade, and I didn't grasp much of it then, as I remember. . . . Theorems, postulates, congruent angles. Seems to be a lot of memorizing and vocabulary."

One week later, her fears were even stronger:

I am experiencing again the close-to-tears frustration I felt at the begin-ning of this course. I could see last week's definition re: an angle repre-sents a part of a circle. I also had no problem with the assignment of two weeks ago: examine the [pattern] blocks and list as many attributes as possible. I have looked over the tangram activity, and read [the articles].
Isosceles right triangle—strikes fear and anxiety.
Trapezoid—
Obviously these words have some meaning to me, and my associa-tions are not coupled with success. I feel like I'm entering a tenuous realm.

Yet, even in her state of panic, Linda recalled her successes earlier in the semester:

Along with this frustration, even before attempting the activities, is the immediate knowledge that I've been here before (this affective state) and have made it through okay.
But I've been avoiding tackling the task, even to the extent of writing this rather than plunging ahead. I'm overwhelmed and I'm going to bed. Tomorrow I'll pick it up again.

Indeed, by the next day, Linda had tackled both her fear and the unknown definitions, and was once again working at the mathematics:

> Well, the Tangram sheet wasn't so bad. A dictionary helped with the basic vocabulary. I now know:
> isosceles—2 equal sides
> equilateral—3 equal sides
> trapezoid—2 parallel, 2 non-parallel sides.
> As I tried the area sheet, the diagrams came quickly, up to #14! Why was I so fearful of these activities? Okay, old patterns die hard.

Having successfully overcome her fears, Linda could now pursue her studies in geometry, playfully reflecting on her errors for what they showed her about mathematics, about learning, and about teaching.

> As my partner and I were examining the [hexagon], I started to think: "OK, angles 120 degrees. To make a bigger [hexagon] the angles would need to be wider"!!!! Augh. This year I've watched my class try to make a bigger square in Logo by enlarging the angle. I know it has 4 equal sides and four 90-degree angles. But that didn't transfer immediately in terms of constant angle size for another shape. It didn't take long, just a few minutes before I realized the error of my thinking, but that error startled me!

And one week later:

> Whoa! Big mistake last week with perimeter. I was assuming that the diagonal of a square was equal to the length of a side! A comment by a peer—"The hypotenuse of a right triangle doesn't equal the other two sides"—set me off on another trail of investigating. So I learned the value of simply measuring, and touched on the idea of the Pythagorean theorem.

While Linda worked on the relationships among the measures of various polygons, she also pondered the nature of mathematics and the implications of ideas that had arisen in group discussion. One reading assignment for the course (Crowley, 1987) contained the phrase "different axiomatic systems," and several members of the class were curious about what this might mean. When Deborah thought about how to provide an example that would make sense in the context of the teachers' experiences, she rejected the idea of referring back to their high school geometry courses. And because the class had not discussed axioms or formal proofs, she decided to compare triangles in two different systems that have familiar physical representations—the plane and the sphere.

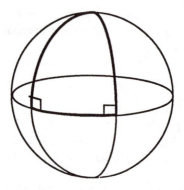

FIGURE 4.6. A triangle on a sphere

"You remember that last week, when we were talking about triangles, you all agreed that the sum of the internal angles is always 180 degrees," Deborah began. "Well, take a look at this." She drew a circle on the board to represent a sphere and drew in an equator and two longitude lines. (See Figure 4.6.) "Can you picture what I have here? Well, if we're on the surface of a sphere we need to change our basic assumption of what a straight line is. Let's say we're going to think of the great circles as lines. In this system we've got a triangle drawn here. Now what about angles?"

The class agreed that the angle formed by the equator and a longitude line must be 90°. That meant that two of the angles already added up to 180° so that whatever the third angle was, the sum of the internal angles would exceed 180°. From this example, the class could see how, starting with a different set of premises—in this case, a different understanding of what is to count as a straight line—different conclusions could be drawn concerning the sum of the internal angles of a triangle.

That night in her journal, Linda wrote about the discussion:

The geometry based on spherical shapes was intriguing. You know how sometimes you hear something and in your gut you know that it's an idea with implications that could be almost limitless? . . . When you drew the globe shape and connected the longitude line to the [equator] and [I saw that the] triangle [had] a sum of angles greater than 180 [degrees]—I knew that as making the absolute most sense of anything I've heard in a long time. . . .But it seems to me that this is a big idea—like Columbus insisting the earth is round. Is my ignorance showing or could this Riemannian geometry someday transform many mathematical paradigms?

Deborah responded to this entry by noting that non-Euclidean geometries have already had a profound impact on 20th century mathematics and physics.

REFLECTING ON THE LEARNING PROCESS

This chapter has illustrated Linda's development as a mathematical thinker. Her first necessary task had been to learn to make meaning for mathematical symbols and operations. For Linda, as for the rest of us, that meant relating those symbols to familiar contexts from daily life. Using those contexts as models, she could negotiate mathematics problems and recognize the logic underlying mathematical patterns.

Once the symbols were grounded in this way, she was able to extend her thought to the next level of abstraction, exploring mathematical relationships, conjecturing and generalizing, without recourse to physical models. And as she evaluated different ways of making meaning, she was able to pose deep questions about the nature of mathematics and about how it differs from other disciplines.

Linda's progress as a strong mathematical thinker was accelerated by her concomitant reflection on her own learning process. By analyzing the elements of that process, she could manage and control her experience in order to best enhance her learning. For example, she realized that when exploring an idea she must look to familiar tools and contexts. "Any meaningful learning must be connected to ideas, knowledge I already have, even if the new 'stuff' forces me to rearrange, change, let go of previously held 'truths.'"

She learned not to be thrown when she saw contradiction, but rather to look for its resolution. With a better understanding of her learning process, she had learned to wait, to endure periods of uncertainty.

The complexities of math are still baffling to me and I certainly didn't expect to have all the tangles unraveled in one short course. But I have learned that little bites of understanding are possible and, for me, the best way to approach mathematics. I'm not nearly as frustrated by my lack of conceptual understanding of math's big ideas.

That's not to say I'm not frustrated when a new math topic is presented! Goldin's (1988) article about affective learning sets really helped me to see my own learning process. I no longer go directly from frustration to anxiety to fear/depression. I can stop and pick up some tools I've learned to use in this course.

Not only did Linda learn to tolerate her frustration, she also came to identify it as a vital component of learning.

The down stages in the learning process—frustration, confusion, despair—precede meaningful learning. Rewards of learning [seem] greatest after periods of blankness, when persistence and struggle pay off.

In the course of a single semester, Linda had developed a sense of herself as a mathematical thinker. She had come to recognize her own authority over mathematical ideas and knew what it felt like to be a mathematician.

> Because of the confidence and new perspective towards problem solving this course has given me, I was able to follow (albeit gingerly) a line of thought that I never would have attempted to attend to before, and my reward was a personal immediate experience in which I was conscious of the living power of mathematical thought—not someone else's account, but mine!

The instructional paradigm championed by influential sectors of the current reform movement presupposes teachers able to navigate fluently the mathematics they are charged with teaching. Yet many teachers have only a tenuous understanding of how mathematical meaning is made. Only as such teachers can be induced, in in-service activities, to engage in explorations of number and space will they begin to infuse content into the symbols and procedures they have known since childhood. And only then will they come to know the importance of working through problems on their own and the sense of enhanced intellectual autonomy that that can bring. Like Linda, they will now be empowered to continue re-constructing for themselves the mathematics they have been teaching, long after their in-service involvement is over.

Educators committed to helping teachers on the journey toward a new instructional practice must recognize that effective teacher development begins with a challenge to teachers' conceptions of how learning takes place and what mathematics is—for teachers can move only by starting out from where they are.

Part II

TRANSFORMING MATHEMATICS INSTRUCTION

5

Stages of Development: Jill Lester

At times nothing was happening; I was at a stuck point. Then suddenly there were places of new understanding, and I'd redefine what my work as a teacher was about. Then the class and I would create a new set of definitions about what should go on in the classroom.

Virginia Bastable
Amherst–Pelham (MA) Regional High School

Stories told in previous chapters have shown teachers constructing new conceptions of learning, teaching, and mathematics content. From experiences of once again being students themselves, they envision their own mathematics classrooms transformed and then begin to plan for specific changes in their teaching practice to help bring themselves and their students into that desired classroom.

However, creating an instructional practice guided by constructivist principles requires a qualitative transformation of virtually every aspect of mathematics teaching. For a few—Jill Lester, the subject of this chapter, is one—this may be relatively easily and rapidly accomplished. Reckoned in absolute terms, it may take six months to a year of hard work, uncertainty, and occasional frustration. But whether the transformation takes six months or three years or more, the process is developmental and comprises a series of discernible stages, each building on the achievements of its precursors while overcoming their limitations in a more embracing and coherent unity (Foster & Pellens, 1986; Hutchinson & Ammon, 1986; Simon & Schifter, 1991). Jill's story is meant to exhibit the principal stages through which teachers pass as they work to make their processes of instructional decision making consistent with their new constructivist commitments.

Jill begins her journey by trying some of the teaching strategies she has been exposed to in the summer institute, but, as with most of her peers, the results of her initial efforts are uneven. Yet she perseveres, and such innovations as group problem solving with manipulatives become routine elements of her practice. However, the introduction of these techniques will not automatically realize the classroom glimpsed in the vision. Instead, decisions about their use will have to be subordinated to an increasingly precise analysis of each student's understandings and consequent learning needs. As this shift is attained, Jill begins, for the first time, to really understand the mathematical principles, the "big ideas," that underlie the curricular topics she is charged with teaching her second graders.

INITIAL CONCEPTIONS OF TEACHING

When Jill, a 38-year-old mother of two, applied to SummerMath for Teachers, she had recently returned to full-time classroom teaching after a long spell as homemaker and substitute teacher. Before she entered the program, Jill's teaching style was fairly conventional. For reading instruction she used basals and workbooks and grouped her students by levels. Her mathematics instruction relied on a standard text; her students proceeded through workbooks and occasionally used manipulatives. In all subjects curricular content was determined by the texts, the lone exception being a language arts journal project that she had begun with a great deal of success.

Jill's *practice* was conventional, but she seemed to have an intuitive sense of what her teaching could become. She simply had no idea how to get there. This was apparent in her application to the SummerMath for Teachers institute, where she was asked to describe a favorite classroom lesson:

One of my favorite math lessons was actually a happy accident. My goal was to introduce the concept of "tens and ones" to my first-grade children. I began the lesson with envelopes of plastic markers for each child. Each envelope contained markers numbering anywhere from 11 to 19. Each child was directed to open his/her envelope and to count out 10 markers, group them, and determine how may were left over. We wrote the numerals on the board and discussed how many tens and how many ones there are in 17 or 14 or 12, and so forth. After about 10 minutes of exploration, a child shouted out, "I'll bet that will work for bigger numbers, too!" We tried it with larger groups of markers, and the children were excited about their "new discovery." In one lesson, most of my 20 children grasped the concept of tens and ones and were able to apply the concept to numbers ranging from 11 to 99.

Clearly Jill was aware of the power of the experience of discovery. She had described a mathematics activity she had devised in which her students had conjectured and verified an important insight into the logic of the base-ten number system. Yet it had only been a "happy accident." Although she could intuitively recognize good teaching when it happened, how to organize her instruction to facilitate such learning on a daily basis was a mystery.

AN EMERGING VISION

During the institute, Jill was an involved and highly motivated participant. By the end of the first week, she had made some astute observations regarding her own learning process. In her first synthesis paper she wrote:

This week has been an experience that will live with me for a long time. It has been challenging. It has been exhilarating. It has been exhausting.

Through the week's program, I have learned a great deal about myself and my own style of learning. I have found myself learning from my own errors. With the understanding of errors in my thinking, I have gained greater understanding of the concepts involved. Thus came the discovery that the correct answer isn't nearly as important as the thought process.

Jill also reflected on the value of peer interaction:

My newest awareness of myself is the need that I have of some private space when I am learning. I find that I am resentful when I have to share my experiences with others before I have come to terms with them on my own. . . . Yet I've also learned that explaining one's though process requires a great deal of understanding. I've learned much from listening to my colleagues and by having to explain my thought processes to them.

Most important, perhaps, Jill had hit upon a fundamental feature of the learning process:

I have noticed that the specific needs of the learner determine what the individual learner might discover from a task. For example, the [mathematics] activity was designed for the whole group, yet each of us interpreted the activity to fit into our individual fields of knowledge. . . . Learning Logo is another example of this. Each day, I approach class with a personal goal. I find that as I attempt to program, new needs arise. Although I am working in a group setting, I have my own agenda. . . . So do children.

By the end of that first week, then, Jill had made some important connections between learning and pedagogy, connections that helped define her vision of good teaching more sharply. Now, into the second week, her thoughts turned to the innovations she hoped to introduce into her primary-grade classroom. In her second synthesis paper, she discussed her experience of the one-on-one interview exercise described in Chapter 2.

This week has taken on a different focus for me. How much of what I
have learned can realistically be assimilated into my teaching style? . . .
The two days that we spent with individual children afforded me a completely safe environment for trying out my new learning. For example, as
the time for teaching approached, I listened to many new ideas. (Why I
had not thought of some of them remains a mystery.) I latched onto the
idea of beginning with meaningful word problems as a stepping-stone for
the work with manipulatives. As I reflect back on last year's teaching, I
can visualize my classroom bustling with activity as the children used manipulatives. Yet I smile when I realize that as soon as I felt that *I* knew the
concept, the manipulatives disappeared. (Didn't I know the concept before I began? How fair was I being to the children?)

Jill's paper went on to assess what she could realistically ask of herself as
the new school year opened, and it is a measure of her self-awareness that she
set clear and achievable initial goals:

I feel, now, that I know how I will try to apply my new learnings to the
classroom. Two modifications are about all I can handle at one time. The
addition of manipulatives (that don't disappear into boxes on the shelves)
will be the easier of the two changes. The second will be to try to give
nonjudgmental feedback to the children during problem-solving sessions.
That will be more difficult and will require lots of practice. I like to smile
and nod my head, though not necessarily in that order.

As the institute drew to a close, Jill's enthusiasm remained high. Since
Cathy Fosnot would be serving as her follow-up consultant in the coming year,
Jill met with her briefly to discuss their work together and to schedule a first,
after-school meeting in September.

FROM VISION TO PRACTICE

Through her work in the summer institute, Jill had begun to develop a
new—if as yet only rudimentary—conception of the learning process, one that
contrasts sharply with the commonly held notion that learning is a process of
accumulating information. Analyzing her own experiences, she had come to
value the opportunity to make mistakes and, by confronting them, to construct
deeper understandings. However, running a classroom according to her developing understanding of constructivism would not prove an entirely straightforward task.

First Faltering Steps

Cathy's initial visit came one month after the institute had ended and one week after Jill had begun meeting with her class. Cathy started that first discussion by asking Jill what had impressed her most during the institute and what she would like help with during the year.

"I just loved it," Jill exclaimed. "I want to do it all. I want to use open-ended questions, watch my 'wait time,' and use manipulatives with problem solving."

As with most teachers, at the beginning of the process Jill's articulated goals were procedural or activity-based. Though she had her vision of a transformed classroom, she did not respond to Cathy's questions by describing the kind of teacher she wanted to become or the mathematics she intended to teach. Nor did she talk about "facilitating invention" or about "making learning meaningful for the children" by starting with their initial conceptions. Jill was now confronted with a classroomful of children, and she needed to decide what to do with them tomorrow, next day, next week, which work sheets to make up and which manipulatives to have on hand. Jill's progress would result from a complex interplay of manageable innovation and reflective analysis, each refining the other.

Cathy had agreed to follow Jill's lead, and when she returned to Jill's school the following week, Jill had already begun her lesson. She had written *17* on the chalkboard and then asked the children to make up a problem that had 17 as its solution. She gave them a great deal of time to think, and then invited answers. The children she called on offered correct solutions, and this seemed to please her.

Jill then distributed a worksheet containing five word problems requiring simple addition or subtraction and asked the children to work in pairs, using Unifix cubes if they desired. After approximately 15 minutes, and though most of the children had not had time to complete all five problems, she asked them to return to their seats to share their solutions as a whole group.

The first child she called on explained his solution to the first problem correctly, so she proceeded to the second problem: *If a party began at 4:00 and ended at 8:00, how long did it last?* Here there was disagreement. Several children had added and decided the answer was 12 hours; others subtracted, arriving at 4. Although Jill asked them to explain how they had arrived at their differing solutions, the children seemed unperturbed by the conflict and waited to be told which of the two was correct. At this point, recalling her commitment to herself to give only nonjudgmental feedback, Jill was unsure what to do next. Because she did not know how to give them back responsibility for working through the logic of the problem to resolve the discrepancy, she became flustered and put off further discussion to the next day. The class moved on to the remaining problems, and, as the children offered correct solutions, Jill's confidence returned.

With only a few minutes left in the period, Cathy saw an opportunity to show Jill how to draw out students' ideas to use as the basis of discussion. With Jill's permission, Cathy wrote $5 + 8$ on the chalkboard and asked the children to think of all the different ways they could get the answer. After allowing them time to think, she asked them to describe the various strategies they had used to come up with their solutions. Patrick explained that he had "counted on from 5" and that his fingers had helped—he had used 8 of them and then stopped.

Ava nodded her head in agreement. "I did it that way, too, Patrick. Only I began with 8 and used just 5 fingers. It's shorter that way."

Here was an opportunity for a challenge that would heighten the discussion and deepen conceptual understanding, so Cathy countered, "But that's a different problem, isn't it? Isn't that $8 + 5$?"

"No, it's the same!" Ava declared adamantly.

"Can you prove that to us? What a neat shortcut that will be, Ava, if your way will always work." Patrick and the other children were intrigued.

Ava thought for a moment, then came to the board and drew a row of 13 lines. Proceeding from left to right, she circled the first 5, and explained that 5 lines were in the circle and 8 were left. Then, in the same way, but this time moving from right to left, she circled 5 again. "See," she said, "it's like taking 3 from the 8 to make 5 into 8. That makes the 8 into a 5."

The class was getting excited. A third child, Kirk, raised his hand. "My way is kind of like your way, Ava—what you drew, I mean. But it's different from Patrick's. I knew that $8 + 5 = 13$ because $7 + 6 = 13$."

"How did that help you?" Cathy probed, looking puzzled and feigning ignorance.

"Well, I took 1 off the 8 to make 7, then I added it to the 5 to make 6. I knew $7 + 6 = 13$, so I didn't need to use my fingers," Kirk announced proudly.

"Hey, I just figured out an even shorter way!" cried Jim, who could not contain his excitement. "Take 2 from the 5 and put it on the 8. That makes 10, and $10 + 3$ is easy." More hands went up now, but time had run out, and Jill had to get the class to gym.

After the children had gone, Cathy and Jill sat down to talk about what had happened. Pleased by the reasoning she had observed, Jill began, "I didn't know Kirk could think that way. And I love the way you played devil's advocate. It really made them think. I can't wait to try it."

They then brainstormed together to find ways of ensuring that the children would take on responsibility for thinking through the logic of problems. The key was to have the children engage in discussion about *their* ideas, demonstrate *their* solutions, and try to convince others. Jill decided to have the group sit in a circle on the carpeted floor so that the children could more easily show one another how they had solved the problems using manipulatives.

Finally, Cathy asked Jill if she thought the children had had enough time to work through *all* the problems on her handout. Jill explained that she knew they hadn't, but that she had been more concerned that some children would finish too quickly and have nothing further to do. After a discussion of the merits of spending more time on fewer problems but encouraging multiple solution strategies, Jill decided that that was what she would do.

Initial Innovations Become Routine

During those first weeks, as she experimented with new strategies, Jill reflected on what was working, what was not, why it wasn't, and what remedies she might try. Soon her lessons were running smoothly. The children had become accustomed to explaining their reasoning, and Jill enjoyed playing devil's advocate. She concentrated on teaching place value and now routinely planned problem-solving sessions coupling mental math with the use of Unifix cubes.

For example, one day the children were working out the sum of $14 + 15$, doing some wonderful thinking. They shared a variety of strategies, such as pretending the 14 was 15, adding $15 + 15$ to get 30, then taking the 1 away to get 29.

Then one of the children volunteered a strategy that her older brother had shown her: "Add $4 + 5$," she explained, "and you get 9. Then add $1 + 1$ and you get 2. That makes 29."

When Jill countered, "I'm confused, Janine: $9 + 2 = 11$. How did you get 29?" Janine and the rest of the class were puzzled.

One of the children suggested a resolution. "I think it works," Jim offered tentatively, "because those are not really ones that Janine added; they're tens: $10 + 4 = 14$; $10 + 5 = 15$. Those are 2 tens." Place value was being constructed. Discussion continued.

During the conference that followed, Jill expressed excitement over Jim's discovery and then went on to report that she no longer used the textbook. Instead she planned each class according to what had happened in the previous one and what she thought the children needed to know. Of this approach to teaching mathematics, she exclaimed, "It's the most exciting thing I've ever tried as a teacher." Jill had succeeded in creating a classroom that made room for the kind of learning she had come to recognize and value.

Shifting the Focus to the Learner

Although Jill *reported* that she planned her teaching in response to the needs of her students (and in many respects this was so), as the weeks wore on she seemed more enamored of her new "approach" than of looking closely at her students' learning. She needed to decenter, to shift focus from herself and her teaching behaviors to the children's conceptual development. Cathy found evidence for this when Jill returned from an in-service workshop in late October wanting to try some new manipulative materials, such as "logic blocks" or "trading chips."

When Cathy asked her why, she replied, "Because I want the challenge."

Intending to shift Jill's attention back to her students, Cathy asked, "How will the materials help the children?"

But Jill persisted, "I would like to do some classification activities with them. The blocks would be a good way, wouldn't they?" She clearly understood the uses to which these materials might be put, but this curricular decision seemed to reflect her own interests rather than the conceptual needs of her students.

At about the same time, Cathy had become concerned that Jill's mental-math activities appeared to have left several children behind. Some of them could not follow the verbal reasoning the others used to explain regrouping, perhaps because the numbers had gotten larger (e.g., 32 + 19). Though Jill was aware that some were having difficulties, she attributed it to their being "slower" or "having a bad day." She hadn't yet realized that by modifying her instruction she could reach all the children.

Deciding to press the issue directly the next time they conferred, Cathy named three children who were obviously lost during the mental-math activity. "I'm concerned about Patrick, Jessie, and Jeanne. Do you think they understood the strategies that the other children were describing?"

Jill admitted that they probably had not.

"Then do you think they will develop an understanding of regrouping via a mental-math approach?"

"I thought so at first," Jill responded pensively. "But I'm not sure now. In the beginning they really seemed to understand, and the approach felt so natural."

"What was the difference between the task then and what you're doing now?"

"I think they could visualize smaller numbers, like 18, without manipulatives," Jill responded after some thought, "but they have no concept of 38 or 72. I'm not even sure they understand that 72 is 7 tens plus 2 ones. Maybe they need more manipulative experience."

In prodding Jill to begin to think about instruction in conceptual terms—rather than in terms of techniques and materials—and in relation to specific students, Cathy had deliberately used a line of questioning she hoped would engender disequilibrium. Jill had fashioned mental math, used with counterexamples and skillful questioning, into a powerful instructional technique. But now, in asking Jill to confront the unmet needs of three of her students, Cathy sought to provoke Jill into relating her use of the technique to those needs. The remainder of the conference time was spent brainstorming for ways to make mental math more responsive to Jeanne, Jessie, and Patrick. Jill decided that she would continue using the activity, encouraging children to take numbers apart and regroup them in various ways, but would have the children sit in a circle to share their ideas, using a set of large, two-dimensional cardboard, base-ten blocks that she would make. She also decided to lengthen her problem-solving sessions to give the children more time to work in their small groups, manipulating numbers concretely using Unifix cubes.

Arriving in Jill's class one day in November, Cathy found Jill and the children already deeply absorbed in the day's mathematics. Some of the children had come up with a series of conjectures that all were involved in trying to prove. These were written on the board: two even numbers added together make an even number; two odd numbers added together make an even number; an odd number divided by an odd number makes an even number; and an odd number plus an even number gives an odd number. The children were working together in groups of threes and fours, using Unifix cubes, attempting to prove or disprove their theories.

When Jill and Cathy met after class, Jill explained how the investigation had come about. During an activity she had designed for exploring place value, some of the children had grouped the cubes by twos in order to count more quickly and had found that, while some numbers could be grouped without remainder, others had a cube left over. Using this experience as a base, Jill introduced the terms "even" and "odd." When the children expressed curiosity about what would happen when the numbers were added, Jill was willing to depart from her agenda in order to let them explore the questions they had defined for themselves. She even saw value in investigating "wrong" theories—like the conjecture that an odd number divided by an odd number yields an even number.

Solidifying the Shift

To encourage Jill to continue her focus on student learning, and to provide her second graders with a place to record their mathematical theories and proofs, Cathy suggested a mathematics journal project. Jill was already using journal writing in language arts and was receptive to the idea that the technique could be adapted for learning across the curriculum. Her own experience of keeping a daily journal during the summer institute had shown her how journal writing could facilitate the reflection process, helping to bring about new insights. The journal would also serve as a space for conversation between teacher and student. Because Jill's second graders were accustomed to writing freely, unconcerned about the niceties of spelling and grammar, and knew to expect her comments in return, the mathematics journal quickly became an integral part of her mathematics program.

In December of that year, Jill wrote an article, "Math Journals: An Individualized Program," for publication in *The Constructivist,* the newsletter of the Association for Constructivist Teaching (see "Resources for Teachers" in the Appendix). Her article not only explains how she was using the journals, but also shows how, at that time, she was thinking about teaching:

> Currently I am teaching grade two. The children are seven and eight years old and have a wide range of ability levels. With the guidance of the SummerMath for Teachers Program at Mount Holyoke College, I am implementing a mathematics program in which the children are "reinventing" math. While I have been quite sure that the children are developing understandings and problem solving techniques that will be of value throughout their lives, I have had nothing concrete to reinforce my theory. . . . At the suggestion of my consultant from the SummerMath for Teachers Program, a math journal was introduced to the children. Our intent was that this would become a vehicle for the children to record their theories and their information in order to prove or disprove their theories. Rather than having the focus be only successful answers, we hoped to emphasize the thinking process behind the answers. However, it turned out to be an even more valuable tool. A truly individualized, constructivistic math program has evolved. The journal entries . . . indicate the facility and comfort with which many of the children are manipulating numbers. They are capable of computing and proving their answers long before they are familiar with the algorithm. In fact, they are constructing their own algorithms and theories. (p. 1)

Jill continues to describe how the journals are used and then offers excerpts from children's entries showing evidence of their thinking. (She corrected the children's spelling for purposes of readability.)

> *Child 1:* 35 + 16. What I did. I took the 35 and I subtracted 5 from it and made 30. Then I subtracted 6 from the 16 and made 10. Then I added the 10 to the 30 and made 40. Then I added the 5 and the 6 and made 51.
> *Child 2:* 35 + 16 = 51. I took the thirty-five and made it thirty. Then I took the sixteen and made it ten. Then I added the thirty and the ten together and got forty. Then I added the five from the thirty-five and the six from the sixteen and got eleven. I made the eleven a ten. I added the ten to the forty and got fifty. Then I added the one from the eleven and got fifty-one. (p. 2)

Jill also provides an example of an exchange between child and teacher which spans several days and in which each responds to the other's entries. Through this dialogue, the child learns to put the knowledge that he has developed about place value into use:

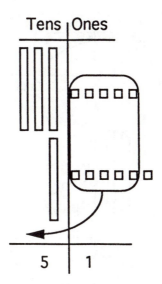

Tens | Ones

FIGURE 5.1. A child's demonstration of 35 + 16 = 51

Child: 35 + 16 = 51. Well, what I did was I added the tens. Then the ones. The 11 I regrouped. I regrouped its ten to the tens place and I got 51. [See Figure 5.1.]

Teacher: Why did you add the tens first?

Child: I did the ones first.

Teacher: Why did you work with the ones first?

Child: That's what I was told to do in first grade. I knew it in second grade because we do mental math. And when we do mental math we add the ones first because it is faster. When we regroup, we need to.

Teacher: I understand a little better now. What happens when you add 46 + 25?

Child: I am learning a lot of math in second grade. I really like mental math. Now I know a lot more. And it is easier and more fun.

Teacher: You are thinking a lot. Please try to add 46 + 25. Draw a picture and explain.

Child: 46 + 25 = 71. What I did was I got the 46 and added the 25. I added the ones first and then I regrouped. And then the tens. I got 71. [See Figure 5.2.] (pp. 3–4)

Using the journals, Jill had found a way to adapt the mental-math activity to meet the needs of individual students. Starting with the child's understandings as evidenced in journal entries, Jill could tailor questions to his specific needs. And she could now assess and plan her instruction according to what she learned about her students' conceptions. Jill's approach to teaching was no longer simply

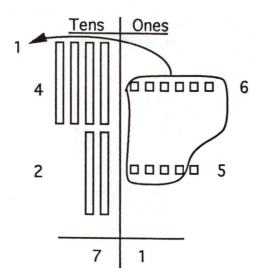

FIGURE 5.2. A child's demonstration of 46 + 25 = 71

an ad hoc composite of strategies, and, if her practice was still evolving, it was increasingly oriented toward understanding and meeting the children's learning needs. Jill's article concludes:

> The addition of a journal component to my math program has served several purposes. The journals have proven to be both a reflection of the child's feelings and a reflection of his/her understanding. Through the journal, the child is receiving individualized "instruction" and the teacher has an invaluable tool for assessing the child's progress and needs. Secondly, the math writing time has fostered a reflection on the process and logic of mathematical reasoning. . . . I believe the journal writing encourages this process. For some children, in fact, the journal might be just the link that makes the discovery of math meaningful. (p. 7)

Finding the Big Ideas for Second Grade

Jill had succeeded in shifting the focus of her instruction from her own actions to the mathematics her students were constructing. Analyzing their understandings and their confusions helped her to determine how best to introduce new concepts, how to structure activities so that *all* students would have access to them, and at what pace to do these things. But as the nature of her instruction changed, she began to reconceive the mathematics content she had been teaching, identifying the central, organizing ideas that are embedded in the second-grade curriculum, but that remain largely hidden from both teachers and students in the traditional classroom.

For example, one day Jill presented the following problem to her students:

Add 6 to me, then take away 3. You will have 11. What am I? After the children had worked on the problem in their groups, Jill asked them to join her to share solutions. Three were offered: 3, 8, and 9. Katy's group had proposed 3, and Jill asked them to describe their strategy first. Katy explained that they had taken 6 and subtracted 3. Kirk seemed puzzled and asked why she had started with 6, pointing out that the 6 was supposed to have been added to some other number. When Katy asked him to explain, he took 8 cubes, added 6 more, then subtracted 3, leaving 11. Several children wanted to know why he had started with 8. But Kirk replied that he hadn't known to, he had just tried different numbers, among them 9. When he discovered that 9 was one too many, he knew the answer had to be 8.

Jill asked if anyone could think of a way to get the answer without just guessing at various numbers until one worked. After much thought, Ava suggested, "If you regroup the 8 with 2 from the 6 you get 10 + 4. Now take 3 away and you have 11 left."

"That's true, Ava," several of her classmates responded with exasperation, "but you still started with 8. How did you know to do that? That's the question!" Ava admitted she didn't know, and the class ended in general perplexity.

When Jill and Cathy met to discuss the lesson, Jill expressed surprise that the problem had proved so difficult. She had assumed that someone would work the problem backwards: start with 11, add 3, and subtract 6.

"Let's look at what they did and what that tells us about their thinking," Cathy suggested.

"Well, none of the children had a systematic approach to the problem," Jill began. "They simply used trial-and-error." After some thought she continued. "No one seemed able to start from 11. They all proceeded by operating on the numbers in the order they were presented, using the operations specified in the problem."

"How is your solution different from theirs?"

"I went backwards. I started at the endpoint and reversed every operation. When the problem said 'subtract,' I added. When it said 'add,' I subtracted. I knew those procedures would reverse what the problem said had been done and that would bring me to my starting point."

Jill's solution relied on her understanding that addition and subtraction are inversely related operations, which was what had to be grasped, if only implicitly, in order to solve this kind of problem in systematic fashion. Though that relationship is taken for granted by most adults—even when they are unable to articulate it—it is a concept that primary-grade students must struggle to construct. Jill went on:

I was thinking about a problem we worked on the other day. I told the children that I had 24 pencils and that I wanted to give one to every child in the class—18. I asked them to figure out how many I would have left.

Although they all solved the problem successfully, they argued about whether the proper operation was addition or subtraction. Some found the missing addend—they started with 18 and added on until they got to 24. They were seeing it as an addition problem. Others solved the problem by starting with 24 and taking away 18—subtraction.

As Jill came to realize that her students were struggling with a fundamental idea—that the operations of addition and subtraction define part–whole relationships in inverse fashion (a developmental leap requiring conceptual reorganization for the young child)—she was increasingly convinced of the importance of confronting their belief that a problem was *either* an addition problem *or* a subtraction problem. For this reason, problems that could be solved using either operation assumed increased significance in her planning.

When conventional texts address the inverse character of the relationship between addition and subtraction at all, they are usually satisfied to counterpose arrays of examples—for instance, *4 + 3 = 7, 7 − 3 = 4,* and so on. But in failing to raise the question of why some problems can be solved using either operation, they underestimate children's ability to successfully mimic the solution procedures they are shown while retaining at some deeper level the belief that the two operations are mutually exclusive. Without the conceptual reorganization that constructing this big idea entails, a defining property of number would remain, at best, only dimly understood.

Jill's prior experience with mathematics, like that of most of her colleagues, had involved manipulating symbols in accordance with rules whose connection to familiar actions and contexts was obscure, if not impenetrable. During this first year she became aware that when her students worked on problems that had meaning for them, their difficulties and confusions, the points at which they needed to step back and take time to ponder, were likely to flag important conceptual-developmental issues, issues that had previously been invisible to her.

Listening to her students' mathematics conceptions, building on their strengths, and identifying their puzzlements, she began to realize that certain themes arose repeatedly. In subsequent years, as different groups of children struggled with those same issues, she was able to recognize which ones were key to their understanding of the number system and to devise problems to facilitate that understanding. Jill was now teaching with a view to the construction of the big ideas of the second-grade curriculum.

A MATURE AND COHERENT PRACTICE

Four and a half years after Jill entered the introductory institute, she enrolled in a SummerMath for Teachers course (Mathematics Process Writing Project; see "1990" in the chronology section of the Appendix) whose purpose was to encourage teachers to write about their own mathematics classrooms. In her paper (Lester, 1991), Jill chose to convey the feel of her instructional process by describing how she and her students (15 boys and 7 girls) had worked on a six-week-long unit on subtraction with regrouping.

Excerpts from her paper afford a window into her classroom, enabling the reader to appreciate how the various strands of insight from that first year of transformation had since been woven into a mature and coherent practice.

> One day, I wrote a word problem that would encourage the children to reach beyond the word "give" when they were thinking about subtraction. That ordinary word problem . . . created an opportunity for the children to make connections and to pose questions that I could neither have planned nor predicted. The problem read: *I went for a ride to Vermont which is fifty-four miles from here. After traveling for twenty-seven miles, I stopped to have a cup of coffee. How much farther did I have to travel to reach Vermont?*
>
> The children spent some time thinking about the problem and tried to solve it mentally. They offered the numbers 34, 27, 33, 30, and 35 as possible solutions.
>
> Tom looked puzzled. He raised his hand and said, "I don't like the answer 35. I'm not sure why it bothers me."
>
> Steve listened to Tom and spoke out, "I know why 35 bothers Tom. You see, 50 − 20 is 30; 4 − 7 is . . ." He paused, looked at the ceiling, shuffled his feet, and blurted out, "I don't know what 4 − 7 is, but I think the answer has to be less than 30."
>
> Alan couldn't contain himself. He stood up, and pointed at the board where I had written "4 − 7 = ?." He said, "I know what 4 − 7 is. It's −3. The numbers go lower and lower."
>
> Tom was not ready to entertain Steve's idea or Alan's idea. He was still engaged in his own pursuit to make sense out of the situation. He said, "I don't know whether to add or subtract. It seems like more of a plus than a minus, but . . ."
>
> Sam raised his hand quickly and began to speak at the same time. He said, "I think it's both. You can count down from 54 to 27, or you can count up from 27 to 54."
>
> I had not yet said anything. The children had been sitting in their seats, turned toward the board, listening to the ideas of their classmates, and adding some of their own thoughts to the process. However, I felt that not all of the children had been able to follow the discussion, and I wanted everyone to share in the explorations. Though I could not predict what the children would choose to explore, I felt that they were beginning to look at connections between addition and subtraction, and to look at the relationships between numbers. So I chose to move the entire group into a circle on the rug where the children could demonstrate their ideas with a large set of cardboard base-ten blocks.
>
> As we moved from our positions in the front of the room to the back of the room where we usually sit to process our math problems, there was a great deal of commotion. The children were muttering and sharing ideas as they went. As they

sat down, I heard Molly say, "Now I'm really confused. I'll have to listen carefully and ask good questions."

Tom begins the discussion by saying, "I think it sounds like a plus, so I'm going to start with 50." As his classmates challenge him, "Why 50?" Tom asks for help and watches as Paul, Molly, Phillip, Mark, and Susan, each contributing suggestions, work to solve the problem. They start with 2 rods and 34 cubes, and remove 27 cubes, leaving 27 behind. When they are done, Tom says, "Now I get it! I was mixed up with the plus and the minus. I was mixed up about where to start."

> Though quiet [as Paul and the others solved the problem], Tom had continued to participate actively in the group. He had listened to the discussion and come to the conclusion that the reason he had difficulty finding a solution to the problem was his confusion about "where to start."
>
> As I listened to Tom, I could feel movement to my left. I had barely turned my head when I heard Molly say, "I'm still wondering if you can say '4 − 7.' "
>
> There was a thoughtful pause. I thought everyone was thinking about Molly's comment about negative numbers, but I quickly learned that I had been mistaken. Everyone had been thinking, but not everyone was focused on the same mathematical ideas. Mark took control of the group and continued on as if no one had even paused to take a breath.
>
> *Mark:* Now, I'll solve the problem by adding. I have 27. (He set up 2 ten-sticks and 7 one-cubes.) I need to find out how many more it takes to make 54. First, I'll take 2 more tens. (Mark added 2 ten-sticks to his arrangement.) That makes 47. Then I need some more ones. I'll take 7 of them, so we'll have 54. (He added 7 ones to his arrangement on the rug.)
> *Jim:* You added up to make 54.
> *Susan:* I can subtract a different way to prove we're right. I'll start with 54.

As Susan starts her solution by laying out 5 rods and 4 units, she introduces a critical step toward the students' understanding of regrouping: she removes 1 rod and starts to count out 10 units to replace it.

> *Molly:* Why are you taking away a ten?
> *Susan:* I'm trading. It's still the same. . . . It's easier for me to trade. That's 54 right there [pointing to what was now 4 rods and 14 units]. I'm going to take away 27.

Jill is confident all along that the class will eventually take this step, the concrete equivalent of regrouping in the subtraction algorithm—if not today, then the next, or the day after that. Since she does not believe it is necessary to single-mindedly focus the children on that strategy, she can allow them to explore other important issues that interest them. On this day, the students begin to explore trading after they feel satisfied that the problem can be solved by either addition or subtraction.

> The children had successfully solved the problem, and I was mentally evaluating my choice of numbers in its construction. I had not considered the confusion that might result from subtracting 27 and having an answer of 27. Sam interrupted my thoughts by rekindling our discussion and taking it in another direction.
>
> *Sam:* I can see multiplication, too: 27 × 2 is 54.
> *Teacher:* Do you agree? (Fifteen children agreed. The rest were unsure. Anna changed her mind twice and laughed about it.) I wonder . . .
> *Paul:* I can divide 54 into two equal parts. (He counted out 4 tens and 14 ones. He divided them into two piles of 2 tens each and then dealt out the 14 ones.) . . .
>
> The children had been working hard for almost an hour, and I could tell that they were beginning to feel tired. I was about to send them back to their seats when Steve smiled, pointed to his head and said, "It's all in here."
>
> Mark crinkled up his nose and laughed. "Everyone was thinking!"
>
> And Sam looked amazed as he stated, "We added, subtracted, multiplied and divided—all with 27 and 54."
>
> The children slowly drifted back to their seats to begin writing, but as they casually strolled they were talking about the mathematics that they were exploring. Math was not over.
>
> This particular day's mathematics lesson was rich in many ways. The children had found multiple solutions to a math problem. They had added on, compared numbers, and regrouped tens into ones to solve the problem. They had discovered relationships between addition and subtraction and between multiplication and division. And they had brought up the idea of negative numbers. They had taken on the responsibility of making a rather ordinary problem into a meaningful experience.
>
> The day's problem-solving session was over, but I was faced with the difficulty of deciding which mathematical ideas I would pursue next. I was excited at the prospect of exploring negative numbers, as Alex and Molly had proposed, and I was excited about the idea of pursuing the exploration of the relationships among addition, subtraction, multiplication, and division. With mixed feelings, I decided to place these ideas "on hold," . . . to pursue the transfer of the children's developing understanding of subtraction to the more traditional paper and pencil task of subtraction with regrouping. (pp. 5–10)

In this one class session, Jill has put into action many of the lessons she had learned. The specific techniques she had decided to try for herself during the introductory institute—starting with meaningful word problems, keeping manipulatives available, and providing nonjudgmental feedback—are in evidence. So, too, are the strategies she developed at the beginning of her year of follow-up: students are given one problem per lesson and are encouraged to find a variety of solutions; lessons begin with mental arithmetic, then manipulatives are used to test and explain ideas; and large two-dimensional base-ten materials are available for whole-group discussion.

Since the routines for mathematics class are familiar, attention can be turned elsewhere: Jill does not need to think about her "approach" and is free to concentrate on what the children are saying, who is speaking up, who is silent, and who may be confused; and the children can attend to the mathematics that is being discussed and the questions it raises for them.

In fact, once the initial word problem is posed, the students determine the direction in which the mathematics discussion will go. They have learned to take responsibility for their own learning; they know they must listen to one another, give voice to their ideas, and pose their own questions.

While Jill takes a back seat in the discussion, she pays careful attention the process. Her interventions are designed to bring in the whole group when the discussion becomes dominated by just a few.

At the same time, she is attuned to the mathematics as it arises in discussion. Though she has decided that the time has come for the class to work on subtraction with regrouping, she is committed "to look for opportunities to explore the mathematical questions that the children [pose] for themselves in their process of discovery—even if those explorations [do] not lead directly to the algorithm" (Lester, 1991, p. 1). Rather than keeping to specific objectives for each lesson, she keeps her eye on the long-range goals she has set for the children.

Teachers who do not understand how children learn mathematics—their conceptual-developmental needs—or what the mathematics is must depend on textbooks to structure their lessons. They have no choice but to assume that the authors know how the content appropriate to their grades must be presented and to believe that, if they march their students through the chapter sequence, the children will learn the mathematics. But when students are taught this way, they are not provided with opportunities to confront and work through the conceptual issues that underlie the math facts and algorithms the texts deploy. No wonder children are bedeviled by the same confusions year after year after year.

By contrast, as Jill becomes adept at identifying the big mathematical ideas she and her students need to address, and as she becomes more confident that her students will learn what they need to know, she is correspondingly less constrained by rigid daily agendas. Her students' discussion of the mathematics of her trip to Vermont shows her that they are on their way to constructing regrouping, so she is not thrown off by their desire to explore how the problem can be solved by either addition or subtraction. And they are, she knows, working on important developmental shifts in their thinking, on another of the big ideas she intends them to construct this year. Then, too, she is pleased to see them begin thinking about other, more "advanced" issues—like the result of subtracting a larger number from a smaller one or the relationships among the *four* basic operations.

At the same time, as Jill allows the class to determine the direction of their mathematics discussions, they are learning important, if unspoken, lessons about the nature of mathematics and their relation to it. They are intrigued by their explorations of how the number system works. They know that their responsibility in class is to understand and that "getting the answer" is merely a means to understanding. They can each identify the experience of understanding and know that when they are confused it is up to them to ask questions. These students have learned that asking questions and proposing solutions that may be incorrect are ways of engaging with mathematics that are at least as respectable and fruitful as computing correct solutions.

Jill's development of an instructional approach based on the new paradigm was smooth, much more so than it is for most teachers. Yet her progress, the result of hard questioning, proceeded through a series of qualitative shifts in her own practice. In the summer institute she had identified key elements of the learning process, and when she returned to her classroom in September she was ready to try out new instructional techniques that would support the kind of learning she was now after. However, once these innovations had become routine, she realized they were not sufficient in and of themselves, but were useful only to the extent that they met her students' learning needs. As she turned her focus from her own behaviors toward analysis of each of her students' understandings, she could plan and modify her lessons to best support their conceptual development—even if that meant departing from the day's agenda. And as the nature of her instruction changed, she began to reconceive the mathematics content she had been teaching, identifying the big ideas of the second-grade curriculum. While initial coursework had helped Jill to create a vision of a transformed mathematics instruction, the bulk of her learning took place in her own classroom with her own students. For it is in this context, where theory and practice interact, that ideas are tested, contradicted, refined, and expanded.

6

The Rug Pulled Out from Under Her: Sherry Sajdak

I have discovered how difficult it can be to explore an idea that has been a misconception. The basis on which it is founded can be so sophisticated that it's not easily undone. Because it has already managed to survive a test of time, it's hard to undo that entire structure or network that has been built around the misconception. . . . The process is as much emotional as it is intellectual, as it calls your own competence into question. There is a period of chaos when a concept is being challenged, knowing that it doesn't fit but not knowing how to make it right.

Rita Horn
Goshen (MA) Center School

Changing your whole understanding and the basis on which you have been teaching for all these years—it's scary!

Jorge Pezo
Warren Harding High School
Bridgeport, CT

Most teachers have spent at least 16 years in the classroom before they enter the profession. All of their experiences in the world of education—both as teachers and as students—have contributed to a deeply entrenched practice of mathematics instruction, incorporating a network of beliefs, often unarticulated, about what should take place in the classroom. But when they encounter an instructional paradigm informed by constructivist views of learning, many teachers find themselves developing beliefs incompatible with their current practice. To the extent they are prepared to confront these contradictions, they must reexamine, or in some cases, perhaps examine for the first time, many of the ideas, methods, and habits that form the basis of that practice.

Instances of teachers coping with the disequilibrium engendered by conflicting sets of ideas and expectations have already been described. For example, in Chapter 2, Betsy Howlett was shown doubting she could ever forsake scripted lesson plans in order to react spontaneously to children pursuing their own mathematical ideas, even as she was coming to believe in the importance of doing so. Also depicted was Pat Collins's struggle to find a balance between emphasis on learners' thinking processes and insistence on correct answers.

For many teachers, the emotional turmoil extends beyond a brief initial period of confusion and self-doubt—it took Sherry Sadjak, the subject of this chapter, two years to regain her confidence. Unlike Jill Lester, who readily incorporated novel ideas into a new and rapidly developing classroom practice, Sherry floundered, suspended between clashing pedagogical visions, unable to let go of the old but equally unable to deny the power of new, albeit confusing, learning experiences.

One early afternoon, 10 months after Sherry had attended the initial-level SummerMath for Teachers Institute, Deborah went to her school to interview her. While receiving weekly follow-up visits from another staff member that spring, Sherry had also been attending the program's mathematics course. As her instructor, Deborah had been concerned initially over Sherry's frustration with herself and, later, over her personal withdrawal. "Before I turn on the tape recorder," Deborah said, once they'd found a place to sit, "I want to tell you that I have a lot of respect for you and, from everything I hear and see, I think you're doing a good job teaching." In response, Sherry began to cry, the tension she'd been feeling over many months finding temporary release.

At the time of the interview, Sherry, a fifth-grade teacher, had been teaching for five years in a small town in western Massachusetts. Her school, grades 5 through 8, is departmentalized and each day she meets with a total of five mathematics and science classes. For the purposes of this book, Sherry's story is especially helpful because she allowed herself to experience so deeply the discomfort she felt with the many new ideas she encountered in her work with us, ideas she could neither smoothly integrate into her established practice nor dismiss out of hand. She talked and wrote eloquently about how the mathematics she was learning was different from everything she had previously been taught; about how her conceptions of what it means to learn were changing; and about the self-doubt she felt as her sense of what should happen in the classroom was challenged.

To repeat once again a point made in earlier chapters, the current wave of mathematics education reform proposes a transformation of teaching practice so profound that it necessarily and radically challenges teachers' sense of professional competence. When Sherry was asked if she was willing to share her story, she replied, "Yes, because I think it will help other teachers. Had I not felt so alone with all of these feelings and failures, I think it wouldn't have felt so bad."

A CLASH OF VISIONS

Teachers typically conceive of the mathematics that they teach as a heterogeneous body of topics, concepts, and procedures for getting answers (Ball, 1989; Thompson, 1984). They do not think of mathematics in terms of formulating problems, making and pursuing conjectures, or evaluating alternative mathematical claims. Instead, for them mathematics instruction essentially involves showing students procedures for getting answers and then monitoring the students as they reproduce those procedures. Instruction may also include explanations of how the teachers understand the techniques in question as well as exercises in which students manipulate physical objects, like beans or blocks, according to teachers' directions.

Thus, when teachers come to a SummerMath for Teachers Institute, many experience mathematics for the first time as an activity of construction, evaluation, and exploration, rather than as a finished body of results to be stored away. And for the first time they sense that mathematics instruction can be an invitation to the exploration of ideas, rather than a laying on of facts, rules, and procedures.

Some teachers enrolled in the institutes choose to resolve this clash of conceptions by dismissing the new possibilities as irrelevant to their own practice. "This is only for the smartest kids, not my students." Or, "This is only for remedial work, but I teach the top track." But most institute participants find that they cannot ignore the depth and intensity of their own learning experiences. However, integrating them into a coherent picture of mathematics instruction means "unlearning" much of the mathematics they already know.

Before entering the SummerMath for Teachers Program, Sherry had a very clear and orderly way of teaching mathematics, taking the assigned textbook as the authority on content and approach:

> We teach the math we teach because it's what you do to get through math. The joy was mastering an objective in a day to go on to the next objective. Many objectives made a unit. The more you taught the better.

Her understanding of the nature of mathematics derived from over 20 years of combined experience as student (Sherry had begun college as a mathematics major) and as teacher in the traditional mold:

> Mathematics is the study of numbers and how they relate to each other and the world around us. Those numbers include time, temperature, weight, and spacial configurations. Mathematics includes symbols and rules for dealing with numbers.

For Sherry, mathematics was a finished sequence of topics in the mathematics textbook. It was not a process; it was not something you discussed; and it did not involve an exchange of ideas. Formulas and rules defined a single correct path to the correct answer for any given problem.

> To me, math was always very ordered and logical, a progression of steps leading to a correct answer. . . . Math was something to be done (not explored; not thought about, just done).

Sherry's decision to enter SummerMath for Teachers came out of a conversation she had with a colleague who had participated in the program. Sherry had been concerned that her students were bored with mathematics and that, two weeks after being evaluated, they didn't seem to remember the material on which they had been tested. In response, her colleague described how, following her participation in the program, mathematics was now exciting to her first graders. So Sherry came, thinking that "SummerMath would give me the bag of tricks to quietly, painlessly perk up teaching math."

What Sherry found instead was a topsy-turvy two-week experience that did not fit either her conception of mathematics or her approach to mathematics teaching. New content, new expectations, new classroom structures made her head buzz.

> Everything was different—from the first day. . . . I worked with Lisa [Yaffee, the subject of the following chapter, on the first set of problems] . . . and I could come up with a formula that I thought made sense to me. . . . Lisa could not do the formula, but she could break everything down and come up with a picture. . . . She could look at that same problem and come up with this . . . , in my mind, very bizarre way of solving it. . . . It was very hard.

For the first time, Sherry was confronted with activities that challenged her restrictive notion of mathematics as manipulation of number. She found it disturbing enough to see alternative approaches, like diagramming, used to solve problems she thought should only be solved by formula, but even more distressing to her was the support program staff gave to these "bizarre" methods. And on top of that, her formulas—to Sherry the "authentically mathematical" means of solution—were not encouraged unless accompanied by an explanation.

> I really think sometimes that I can pull formulas out of my head. They're neat compact formulas and I can plug numbers into them and they work, but I don't have a clue as to why they work. That used to be enough, but it's not enough [here], and that was hard for me.

Challenged in new ways, Sherry found herself nearly infuriated by her inability to explain familiar mathematical "truths."

> I think I should know all this. It sounds vaguely familiar, but even though I know something (like dividing by 0 is meaningless or undefined) I can't find a reason to explain it or an example to demonstrate it. . . . I realize I know things because they are in my mind, but I don't know how to explain the "whys."

Sherry struggled not only with a very narrow conception of the nature of mathematics, but also with a limited notion of what it meant to do mathematics well. Worse yet, she felt that her own performance did not measure up to her criteria. For example, while many teachers derived from the Xmania activity the realization that physical objects are powerful tools for making sense of mathematics, Sherry was angry with herself for remaining dependent on the blocks. "Okay," she told Deborah, "I was willing to work with them at the beginning. But by now I should be able to just do the math. I should be able to just do it on paper. But I can't!"

Sherry could have resolved her disequilibrium by rejecting everything presented in the institute—methods, expectations, mathematics content—as wrong or inappropriate. She could have said that none of this applied to her charge to teach fifth-grade students to add, subtract, multiply, and divide. She could have, but she didn't—for in spite of her frustration and confusion, she was not willing to ignore all that was happening around and inside her.

Sherry had inklings of the difference between getting an answer and being able to explain it. She seemed to recognize that even with all her semesters of college mathematics, she did not "understand" the problems as well as Lisa who, though fearful and anxious, applied her own informal logic. Soon she began going after that understanding, asking her own questions: "Why does it make sense to add when you subtract a negative number?" "What connection is there between my formula and Lisa's picture?"

Although she was as yet unable to articulate the difference between getting answers and explaining them, Sherry sensed that the distinction was somehow related to the dissatisfaction that had led her to enter the program. Watching videotapes of children who could get correct answers but who, when questioned by interviewers, revealed serious misconceptions of what they had just done, Sherry knew her students would fare similarly in the face of such probing. And she hoped that the instructional approach modeled in the summer institute would help her students remember what she had so recently taught them. Thus, as she got ready for school in September, she was resolved, though with considerable trepidation, to try some of the techniques she had encountered in the institute, especially small-group problem solving.

Like Jill Lester, Sherry would begin the year with specific procedural innovations. But where Jill was predisposed to an alternative teaching practice if only she could figure out how to put one together, Sherry had no intention of fundamentally transforming her classroom. She thought she could arrive at a "happy medium" by adding a small-group problem-solving session once or twice a week.

Had Sherry's intention been merely to introduce some of these techniques in order to "perk things up," perhaps she would have found that happy medium. But her goal for her students was the achievement of an as yet poorly defined sense of greater understanding. What Sherry did not see at the time was that her evolving expectation of more substantial learning would prove largely incompatible with her past practice. "There seems to be no middle of the road," she later said. But that gets us ahead of our story.

A YEAR OF QUESTIONS

A mathematics instruction consistent with the new paradigm requires a radically different sense of the discipline than most teachers now possess. David Cohen (Cohen et al., 1990) of the National Center for Research on Teacher Learning at Michigan State University has described this dilemma:

> [Teachers] would have to acquire a new way of thinking about mathematics, and a new approach to learning it. They would also have to additionally cultivate strategies of problem solving that seem to be quite unusual. They would have to learn to treat mathematical knowledge as something that is constructed, tested, and explored, rather than as something they broadcast, and students accept and accumulate. Finally, they would have to un-learn the mathematics they have known. Though mechanical and often naive, that knowledge is well-settled, and has worked in their classes, sometimes for decades. (p. 93)

One of the implications of the analysis is that teachers must embark on two formidable projects at once—they must learn a new version of mathematics while unlearning the old, even as they replace familiar ways of teaching mathematics with new approaches. And all the while, they remain responsible for learning in the classroom and accountable to parents and school administrators. No wonder the stresses can be intense.

Sherry started out the year intending to introduce some new techniques she'd encountered in the institute:

> I had some little bits and pieces, and every now and then I'd try them in the classroom. . . . [I'd put] kids in groups and work on things. That sounds like a small piece, but it wasn't.

However, her initial attempts were generally unsuccessful. Given a sheet of computations to do and despite her instructions that they cooperate, Sherry's students tended to work alone or spend their time socializing. Had she been a less reflective teacher, she could have decided, after these first halting attempts, that her students simply would not work in groups, ending her struggle right then. But she *was* reflective, and concerned about her students' understanding. So she persisted:

I found certain things didn't work well in groups, [like] a sheet of addition, subtraction, multiplication. What purpose did that serve to do in a group? So . . . [I] had to come up with something else. All right, [try] word problems. . . . But most of the word problems that I was finding were straightforward; they weren't challenging, so that the kids knew them anyway. Then I had to start looking around for things that would challenge them.

This process, she emphasized, took several months. "This sounds like it happened relatively quickly, but it didn't."

While Sherry experimented in this way once a week or so, the rest of her class time was used to pursue her regular mathematics program. But she was finding it less and less satisfactory.

[I was] still giving them those long-division problems to do, but yet trying little things from SummerMath. This stuff that I had [previously] been trained to do wasn't very fulfilling for me and wasn't very fulfilling for the kids. And I still didn't quite know what I was doing in SummerMath so I was always at this point of frustration. . . . If I could just get it. If I could get something better than I was doing. And I wasn't sure what it was or what I was looking for.

In early spring, when Sherry began to teach fractions, she was able to find the type of problem that was appropriate for group work. By then—Sherry had now had three months of classroom follow-up and was enrolled in the program's mathematics course—her follow-up consultant entered a classroom in which there were many centers of activity, with students working on a sheet of fractions problems in groups of threes and fours. Some children handled colored plastic rectangles, others drew pictures, still others did calculations. They listened to one another as they explained their diagrams or demonstrated what their plastic pieces showed. Sherry moved from group to group, looking for signs that a child here, a group there, were stuck, staring down at their fraction sheet.

After about 30 minutes, Sherry called the class together and, having visited each of the small groups, selected two to show their solutions to a particular problem.

"So, look at this," Sherry said after the second group had finished their presentation. "Vivian's group got an answer of $\frac{5}{16}$. Simon's group got $\frac{3}{8}$. How can that be? Who's right?"

Viewing all this, her follow-up consultant reported that it might have appeared that Sherry had succeeded in realizing a new mathematics pedagogy. After all, her students were solving problems in small groups. She had found problems that were engaging and challenging to them. They used manipulatives and diagrams as well as conventional symbols to represent fractions. There was a rich variety of thinking in the class, and Sherry had even learned to use conflict to promote discussion.

Yet, in private, Sherry was still troubled by many questions. Though she had been able to make all these changes with help from her classroom consultant, and had reason to think that her students would understand fractions better now than in the past, she still didn't know just what she wanted them to understand, what it meant for them to understand, or how that understanding would come about.

What are the topics I should teach? she wondered. Having always implicitly trusted the judgment of textbook authors, she now thought, for the first time in her life, that what they offered simply didn't make sense.

> We do (used to do) next three equivalent fractions. [For example, given ⅓, the answer is ⅔, ⅜, ⁴⁄₁₂.] When I teach through the context of word problems there is no need for that information. . . . Is a fraction incorrect if it's not in lowest terms? I always thought so. . . . Well, I don't think it is, all of a sudden. So why we do spend 42 million years on simplifying fractions?

If the textbook could no longer be relied upon, how was Sherry to know what she should be teaching?

What is problem solving for? "Are the problems the tool for understanding fractions or is the object to master problem solving?"

One of the problems on her fractions sheet had stimulated a lively discussion—*Joe's mom bought ½ gallon of ice cream. Joe ate ¼ of the container. How much was left for his sister?* The class offered various ways of looking at it, including both numerical calculations and diagrams. But Sherry was left wondering just what she wanted her students to learn from it.

"Some students can solve this by equations," she observed, "and everyone can look at a picture and *see* the answer." But what was the point? "If they can solve multiplication and division word problems with diagrams, does that mean they don't have to learn how to multiply and divide?"

"I have to work hard to follow their explanations," she said.

Sherry had never had to listen to complex explanations from her students before. She had always known when her students gave her a correct answer. But when "one student found a rule to determine, given two unequal fractions, whether one fraction was larger or smaller than the other, he had to explain it several times to me before I understood it. I resort to math—finding common denominators."

And, she asked, *what is mathematics, anyway?* Even as Sherry was changing her instructional style in these ways, with her students working on engaging problems and discussing and debating their solutions, she was not yet convinced they were doing math. True, she could appreciate the validity of her student's explanation of how to find the larger fraction, but she did not yet consider what he had done "mathematics." Somehow with all of her doubts, that was still the formula in the book.

In fact, Sherry was developing a new understanding of mathematics at the same time that she was learning a new way of teaching it. In the process, deep questions were being raised for her, and, though she discussed them with her classroom consultant, ultimately she alone had to construct the answers that made sense to her. Meanwhile, she was being called upon to make instructional decisions that required an understanding of this "new version" of mathematics that she didn't yet possess. Simultaneously, her old conception of mathematics was eroding, along with her old instructional practice:

> The way I taught math doesn't make sense anymore. The textbook (a new one, that *I* insisted on [a year ago]) doesn't make sense. I'm still feeling like the rug has been pulled out from under me because I know what I want to change but not quite how to do it.

Sherry now fully recognized the futility of her initial hope that a painless compromise could be struck between her old conceptions and the new ideas with which she now wrestled.

> I guess it's awesome to get the chance to look at mathematics so differently. It's also overwhelming. There seems to be no middle of the road. It's not enough to throw in some word problems once or twice a week. (I thought I could create the happy medium! *Ha!*) If nothing else I realize how useless most things are out of context. How nothing makes sense mathematically if there's no concrete picture (even in one's mind) of what it is we're talking about.

Near the end of the school year, Sherry could see that she was still in the middle of a process that she had to trust would eventually work itself out.

> What I have learned about my own process . . . is that I learn *slowly* and I want to learn quickly. I keep hearing that this method . . . builds confidence, but I never feel confident. I hope I'm learning, but then I just think I haven't learned enough. But then maybe it's like skiing. The hill always looks so long and steep until you ski it a few (more than a few!) times.

RESOLVING DISEQUILIBRIUM

After another year in the classroom and an advanced-level summer institute, Sherry again met with Deborah to talk about her progress and how she felt about it. Sherry's tone as she began was completely different:

The advanced institute brought everything together for me. All of a sudden all of the pieces sort of fit together—the teaching, the modeling of the teaching, the subject matter to be presented, a lot of questions I had about learning. I mean, I still have the questions, but . . . it's like all these things that were floating around somehow have been gathered together in a basket. And they make a nice arrangement. I don't have all the answers, but there's something that makes it all come together.

Simply having more time to reflect, to sort out what works, what doesn't, and why, undoubtedly helped Sherry to resolve some of her conflicts and develop a more coherent perspective. Yet in analyzing her own process of change, Sherry described events that suggest that more was involved than just the passage of time—they illustrate instead Sherry's developing recognition of her own authority and control over her practice.

During the first year after the introductory institute, Sherry was still trying to integrate the "rules for good teaching" she had learned in college with what she thought SummerMath for Teachers was prescribing for her. Relief first came when, during the spring of that year, program staff "gave her permission" to ease up on herself.

[When I was feeling so overwhelmed, Deborah] said to take one piece, one class, in particular. . . . I picked my favorite class, the one I thought would do the best and work well together, and I set myself up for success. When I did that, I found that the other classes—I couldn't ignore them. And some of the things I learned with this other class I could then use. . . . So, it's interesting that . . . focusing on one class, I think . . . benefitted all the classes. It somehow unpressured me. I wasn't so . . . fragmented.

Much of Sherry's difficulty in creating a coherent set of ideas to guide her teaching practice stemmed from her desire to "do it right." We have subsequently witnessed this paradox many times in our work with teachers. Their attempts to act out what they think we want them to do interfere with our goal for them: to construct a conception of teaching that addresses the experiences we provide but is based on goals for learning that they formulate for themselves, that make sense to them.

Most participants' previous teacher education courses were designed to convey, frequently through lecture, sometimes through demonstration, specific behaviors to be enacted in the classroom. Thus, when teachers enter SummerMath for Teachers, they do so with the comparable expectation that they are there to be exposed to the "latest" fashions in classroom technique. And though our intentions for them are quite different, they conclude that because we use certain instructional techniques, we are telling them that they must use these techniques, too—that that is the purpose of our work. For example, Sherry believed that "if I wasn't doing word problems in every single class every single day, working in groups, then I wasn't doing SummerMath."

So it was when she gave herself permission to reject the techniques she had learned in the institute that Sherry discovered what *she* believed and wanted to do.

> One day I said, "I'm never doing SummerMath again. I'm absolutely going to refuse to do it, I don't care." . . . So I slammed the door, and it was like, Oh, I freed myself from that. Now what is it that I want to do?

As Sherry pondered this question, she was surprised to discover that she wanted to use the very same strategies in her mathematics instruction. Reflecting on this paradoxical outcome, she realized that she should "not try to think of what you want, but what I want." Now, focused on *her* goals and *her* ideas, Sherry saw that she had changed—that she had begun to think about mathematics and mathematics teaching in a new way. "And that was a revelation."

Having "received permission" from others and, more importantly, from herself to accept or reject new teaching strategies, Sherry had next to release herself from the power of internalized images of authority—authors of textbooks, her own mathematics teachers, SummerMath for Teachers staff—in order to assume responsibility for her own learning. In her interview, she described how she had felt at the beginning of the advanced institute.

> [It was like] a year and a half ago at the beginning of the spring math course. [I had] the same feeling, that I'll never get this. I'm really stupid. What am I doing here? They're going to find out. I'll never get this. Deborah will find out. SummerMath will find out. . . . This summer I carpooled with Janine and the first few days we talked about this. . . . And Janine kept saying, "But SummerMath always takes you where you are, whether you're four feet in front or four feet behind or right in the middle. It's perfectly okay." And she said that almost every day to me as we drove in. "It's okay to be wherever you are, and you may get farther, you may go back." But I said, "How can that be? How can you say that?" . . . So we talked about this every morning. And finally I said to myself, I hear. It's okay. And when I could allow myself to be wherever I was and stop worrying about that part, I think I started really making progress.

As long as approval from instructional staff remained Sherry's primary concern, her own thoughts and beliefs remained hidden. Now, realizing that approval was beside the point, she said,

I could finally take risks. I could finally put all my discomfort aside and say, It doesn't matter. They're not going to find out that you don't belong here because it doesn't matter. So you can do whatever you want to do. It took a long time for me to understand that.

Having come to see that it was up to her to "arrange her own basket," Sherry decided that she could now discard whatever we had offered her that didn't contribute to the arrangement *she* envisioned.

A SHIFT TO THE LEARNING PROCESS

As Jill Lester's story suggests, the process of change often begins with an initial and one-sided emphasis on teaching strategies and behaviors, the shift to an instructional practice centered on student learning and mathematics content being accomplished, if at all, only subsequently. Up to this point in Sherry's development, her self-consciousness and concern for approval had kept her preoccupied with her own behaviors, on how to do whatever she thought she was supposed to do. Once freed from the imperative of "having to do it right," Sherry was able to concentrate on the learning process, beginning with her own. For example, reflecting on fresh experiences in the advanced institute's mathematics sessions, Sherry wrote in her journal:

When I have to explain my thinking, my ideas or my solutions, I am helping myself to understand because I have to clarify my thoughts and express them in a way that helps someone else to know what I mean. This conversation further strengthens where *I* am because of the thinking *I* have to do to express myself. It's not just seeing the solution with my eyes but "seeing" it with my brain, analyzing the solution to express it. The act of talking about the problem deepens the meaning for me. If I'm not expressing myself clearly enough my group asks questions. I need to rethink something to express it in a way that makes sense. Again it's that thinking that must take place for me to communicate to my group. . . . Sometimes listening to someone else's language or description of a solution helps me again to see the problem. I begin "seeing" something with my eyes, my ears, my brain, and then make more connections. This process engages the whole learning me!

Two years earlier, Sherry had been captured by a vague but compelling sense that she wanted to teach for "understanding." At the same time, she was exasperated by the idea that there was more to mathematics than finding correct answers to textbook examples. Now, having achieved the satisfaction of deeper understanding, she could analyze what it felt like and describe how she arrived there.

> What I have come to understand is that it takes a lot of experiences to know and understand something. I must work at finding the solution and I must discuss my ideas and methods with others. This is not because I'm "dumb" and can't do it myself, but because I need to explain and reexamine and verbalize and analyze my ideas so that they are sound philosophies and belong to me.

In these journal excerpts, Sherry emphasizes the importance of explaining her ideas and examining them from a variety of perspectives—something the small-group format helped her do. Returning to the classroom next fall, she would have these experiences and reflections to guide her practice. "SummerMath taught me to do it this way" would no longer be her reason for small-group work. Rather, her own sense of what such work can achieve—her understanding of what kind of process she would be trying to encourage—would dictate when and how to use the approach.

Another of the advanced institute's exercises—a tutorial conducted by secondary teachers for one to three elementary teachers—brought into striking relief for Sherry the superficiality of the understanding her old style of teaching supported. The purpose of this activity is to provide secondary teachers with a safe environment in which to experiment with new strategies, receiving feedback on them, while affording elementary teachers an opportunity to critically analyze the effectiveness of those strategies. In Sherry's tutorial, the secondary teacher who was assigned to conduct four lessons on probability for three elementary teachers initially chose a traditional, expository instructional approach. Sherry wrote:

> His style of teaching was primarily lecture. He started by giving us a textbook definition [and] would ask us questions. I could answer most of the questions by referring to the definition or repeating something that he had told us. . . . [Even though I was completely confused] he thought I was less confused because I seemed to be able to come up with some correct answers. . . . [But] the fact that I could manipulate the numbers in a seemingly meaningful way did not mean that I had an understanding of the material.

In contrast to her expectations for herself in the introductory institute two years earlier, Sherry was now no longer satisfied with getting correct answers. "Knowing and understanding are not listening to someone else explain something, nor [are they] parroting back information to someone," she insisted.

From this vantage point, Sherry could now analyze what had been so unsatisfactory in her teaching practice before she entered the program.

> There were too many times that children seemed to follow the motions but did not retain anything. I think they really didn't know and understand the things I assumed they did. There were usually several students who could repeat what I said in their own words. I thought this meant understanding. I could never figure out why their understanding didn't seem to last very long.

Animating this discussion is Sherry's realization that new understandings of learning, teaching, and subject matter are—and must be—coordinated. She no longer saw mathematics as a set of procedures for getting answers, but as an active process of discussion and debate, of making and communicating sense.

However, this did not mean that Sherry was done. The coordination of ideas had created, as she said, a "nicely arranged basket." But there were still many flowers that didn't yet fit.

For example, she still found herself disconcerted by the idea that a single calculation might, depending on context, yield different answers. Thus, in her journal Sherry addressed a lesson in which institute participants made up word problems that could be solved with the computation $32 \div 5$, but for which the answers were 6, 7, 6.4, and so on—for example: 32 children going on a field trip, 5 to a car, would need 7 cars; but if 32 feet of wood is available for shelves, each 5 feet long, there is only enough wood for 6 shelves.

> It's really hard for me to let go of the "right" answer. The confusion, if it is that, is that the answer depends on the context of the problem. The question, or context of the question, can lead you to the "wrong" answer, because it's not wrong based on the problem but on the straight computation. I realize that this is the point of teaching in context. But I also worry about this. It makes perfect sense to me and is in complete disagreement with the "old" way of "only" one right answer. . . . Do you see what happens? I can't quite convince myself! This new information clearly should replace the old, but I'm having trouble letting go of the old way, even though I fully agree with the new way.

In fact, there were probably aspects of Sherry's "old way" that still made pedagogical sense. Her task now was to sort through what she had recently learned and what she had always done in the past, and to figure out how to create a coherent practice based on beliefs that she herself had actively tested.

As she wrote about the questions she still had and the confusion she still felt, she no longer expressed the same dismay or distress. Returning to Rita

Horn's words in the epigraph—"The process is as much emotional as it is intellectual, as it calls your own competence into question"—we realize that it, too, is a contextual truth. Once Sherry was able to lay to rest the question of her own competence, she could then see disequilibrium as a necessary and vital part of the learning process.

> I have experienced that feeling of frustration and aggravation when I can't seem to get it. I have also experienced the exhilaration and empowerment that comes with knowing. It's such a good feeling that I think it is worth the long and difficult journey we all must take to get there. I want my students to feel that same exhilaration and empowerment.

With an understanding of the process of change and a sense of where in that process she found herself, Sherry was no longer so impatient. She could trust that she would continue to grow, to watch her teaching develop, and, in that, find satisfaction.

> There's another piece of information that I found somewhere—that it takes three to four years to change your teaching style. And I thought, I guess I'm not doing so badly. When I started SummerMath, I knew where I wanted to be and I knew that I wasn't . . . anywhere near. . . . That was two years ago. . . . [Now] I'm closer to where I want to be. And it's okay that I'm not there.

Reading this last comment, one may well wonder what Sherry understands by "there." If her goal is a finished repertoire of behaviors that, once achieved, will become routine, then she will be disappointed, not to say bored. But happily, teaching as Sherry is learning to teach is necessarily disruptive of routine, if for no other reason than that her students will continually surprise her with their own discoveries. "There" is not a point of arrival, but rather a path that leads on to further growth and change. On the other hand, if Sherry's goal is the achievement of a confident yet open-ended practice grounded in her new perspectives on teaching and learning, then she is securely on the way "there."

7
What's So Special About Math? Lisa Yaffee

Participating in SummerMath for Teachers has changed almost every aspect of the classroom curriculum—not just math!

Donna Natowich
Green Street School
Brattleboro, VT

Several years ago I became involved in Whole Language and Process Writing and worked very hard to change the way I taught language arts. When I did that, I found that it changed the way I taught almost everything—everything except math.

Nancy Lawrence
Wolf Swamp Road School
Longmeadow, MA

As their mathematics instruction begins to reflect the influence of constructivist principles, many teachers report that their practice in other disciplinary areas has been affected as well. "It has altered the way I think about teaching." "I ask different kinds of questions now." "I'm really listening to my students." "It has even changed the way I deal with discipline!"

Significantly, however, teachers who develop a constructivist-oriented classroom practice in writing, say, or social studies, or the sciences, typically find their approach to mathematics unaffected by analogous innovations. This is often true despite the acute dissatisfaction they may come to feel with their mathematics work. For example, Pat Collins (Chapter 2), who had done work in whole language instruction, wrote at the end of the institute:

> I'm not experiencing a dramatic difference in what I believed was important for real learning to take place. I believed that people need to construct meaning for themselves for clear understanding. . . . What has changed significantly for me is that I now see more clearly that my . . . methodology did not match my belief about how kids learn. I was aware that math was not like the other instruction going on in my room, but unclear what the problem really was.

The inability of teachers to translate alternative pedagogical approaches into their mathematics instruction likely stems from several related causes: first, they share the generally prevalent conception of mathematics, in which inquiry, exploration, and vigorous debate play little or no role; second, those charged with teaching mathematics do not themselves understand the material well enough even to conceive of an alternative practice; and third, most teachers do not have enough confidence in themselves as mathematical thinkers to allow themselves to contemplate a shift toward teaching for a qualitatively different kind of understanding. Additional discouragement may come because students' habits and preconceptions—formed in the culture of the traditional mathematics classroom and reflecting again the larger society's attitude toward mathematics—are often highly resistant to the changes such a shift implies.

Lisa Yaffee, the subject of this chapter, is not typical of the teachers who enroll in SummerMath for Teachers. After college she spent a year in El Paso, Texas, as a VISTA volunteer, later wrote for a feminist newsletter at MIT, where she worked as a secretary, and later still worked with emotionally disturbed and learning-disabled students aged 8 to 22. She attended Bank Street College for her pre-service education and at age 35, when she entered the program, had been teaching for only a year. Lisa is more outspoken than most, more articulate than most, and willing to challenge others as well as herself.

Yet, though she is not typical, and precisely because she is articulate and outspoken, her story shows what can happen to a teacher with a strong appreciation of constructivism and a real commitment to independent thinking when she changes her ideas about mathematics, and as her confidence in her ability to understand it grows.

LISA'S BACKGROUND

Even before entering SummerMath for Teachers, Lisa envisioned a classroom in which students' active involvement was both the goal and the instrument of learning. Her very reason for becoming a teacher was to help her students develop strong voices. "I came to teaching as a form of social action," she explained, "one which would be more enabling for kids than organizing their parents politically." Thus, her conception of classroom process was already a sharp departure from the traditional classroom in which the teacher tells her students what they should know. "I had an image of [the] teacher as street-fighter, [social activist], not as answerman."

At Bank Street, Lisa received a foundation in pedagogical theory that she described as "developmental/constructivist/Piagetian." She had chosen Bank Street because of its emphasis on pre-service classroom experience. She was surprised to find that "most [courses] were good!"—the exception was an acutely disappointing mathematics methods course. "[It] was the least satisfying . . . course I took. . . . Although there was plenty of 'hands-on' . . . there wasn't enough minds-on for my taste."

Once Lisa began teaching, she was happy with her programs in language arts, science, and social studies, but was concerned that her mathematics instruction was not meeting her fifth and sixth graders' needs. "I didn't teach any math," Lisa later explained:

> I'm embarrassed to admit this but it's true. . . . I just sort of handed the kids a math book. I gave them a placement test at the beginning of the year. I let them work at their own pace and it was pretty individualized and that's how math is being taught in most of this building, from what I could observe. I knew it wasn't working. I felt awful about that. . . . But I didn't know what else to do. I really felt strongly that I didn't know how to teach math.

Lisa was quite aware, however, that her problems had not begun at Bank Street. Like Linda Sarage (Chapter 4), she looked back to the very beginnings of her formal schooling to locate the source of her difficulties. As far back as she could remember, she had been an unsuccessful mathematics student, having learned how to "get by," even as she understood that she was missing the essence.

> Learning [math] from first grade on always followed a pattern: observe teacher or kids, record in the memory or on paper, see what others have recorded, respond. I experienced a certain amount of disequilibrium as I learned how to please, but mostly I assimilated and accommodated . . . everything and everyone. . . . Math was a language I couldn't master. I knew the vocabulary but didn't understand what it meant or how to apply it. I did my best to fake the pattern here.

It was generally believed—by teachers, by family, by Lisa herself—that she was among those people who "do not have a math mind."

> I allowed myself to accept what the "objective" results of tests and teacher response to my muddled efforts at math suggested, namely that I couldn't understand it and never would. . . . As an artist, [loosely] defined . . . , I was missing out on a thrilling set of descriptors.

GETTING STARTED

During her first year of teaching, Lisa attended a series of four afternoon workshops conducted by former SummerMath for Teachers participants where she had no trouble recognizing an approach to mathematics instruction consistent with her beliefs about what education should be. She saw that mathematics could be taught through problem solving using manipulatives, so for the rest of that year she included problem-solving sessions in her mathematics instruction. But Lisa was acutely aware that she did not know how to integrate this work with her charge of teaching arithmetic, geometry, and other mathematics content. Thus, in her program application, she presented a fairly well-defined statement of what she wanted to learn:

1. How do you organize your math program around problem solving?
2. How do you introduce new concepts?
3. How do you structure problems so that previous knowledge of algorithms doesn't block hands-on thinking?
4. How do you teach abstract concepts like exponents and division of fractions with manipulatives?

She also saw that in order to teach effectively, she, too, needed to learn mathematics.

I'm hoping to gain a greater conceptual understanding of math processes, of math as a language and of the fearful symmetry underlying the whole mess. I hope to acquire some confidence in [my] ability to think my way out of a mathematical bag, as it were, so I can encourage and nurture that kind of confidence for my kids. . . . I'd like to feel convinced that exponents *do* relate to me personally (beyond the third dimension). In addition to getting some answers to my questions, I'm looking forward to developing the capacity to ask new ones. Please let me come!

Lisa did attend the institute, but exposure to new possibilities aroused conflicting feelings. On the one hand, she knew that she wanted to change and she saw that SummerMath for Teachers offered an approach consistent with her beliefs. On the other, as much as Lisa was poised to develop a new mathematics teaching practice, her glimpse of the big picture frightened her. Primed to react to a cascade of new insights but not quite knowing how to do so, she raged, furious at the institute and the staff:

What the institute has done is to take away our sense of control. We're supposed to live in the dorm. If we don't, we're missing something. We have our meals and classes at designated times. We do things we hate. We're bombarded with alien stimuli until we're exhausted. These experiences do not compute and we get no processing time. In short, we've been thrown into chaos. . . .
   I could go on and on. Instead I'll summarize. This week I learned what it feels like to be a pre-operational [or] concrete-operational kid con-

tending with formal-operational expectations. It's hard, frustrating, overwhelming, infuriating, demeaning, terrifying and depressing.

Yet, despite her rage, Lisa was developing a sense of how constructivist theory might apply to mathematics instruction. In her synthesis paper she argued that:

> You can teach kids how to manipulate material without "teaching" them understanding. It means that you can't "teach" understanding because the way each person understands is arrived at uniquely. . . . It also means that the teacher has to set up conditions under which every student arrives. Somehow the kids have to be able to structure a problem or concept so they can understand it. Your construct probably won't help.

However, the development of a pedagogical vision, or the refinement of one, was not what Lisa felt she needed at the time. She wanted specific answers to specific questions and at the end of the institute she felt no better equipped to act on new understandings of how mathematics is learned than she had when the session began. In a paper discussing the implications for the classroom of her current theory of learning, she wrote:

> I can answer superficially and cosmetically, according to what teaching should look like, but I don't know how to achieve the effect. I'm not sure how to set up an environment in which kids really learn math, as opposed to one in which they learn how to manipulate materials. It is frustrating to be this stupid, to leave the institute bearing the same burden of needs I carried in with me. I still don't know how to build a math curriculum around hands-on problem solving or how to think mathematically (whatever that is). I still don't understand how to use manipulatives to teach abstract concepts like fractions. On top of the old agenda is grafted a new one: how to help kids refine cumbersome or confusing notation systems which might impede learning, how to probe so that students refine or discard ill-conceived theories in which magic or ignorance account for what's happening mathematically; and lastly, how to show the connection between what students "know," or *really* understand, and what they're supposed to know, namely the abstract operations with algorithms on which they'll be tested and by which results their positions in society will ultimately (and criminally) be determined.

Despite her rage, Lisa had found the important questions. She was frightened by her new knowledge and angry at the implied expectation that it was up to her to figure out how to "do it." But now she needed to find the breathing space and the confidence to begin to construct answers—in her own classroom and with her own students.

With some trepidation, Deborah visited Lisa's classroom for the first time in mid-September following the institute. Arriving early, she observed the last 10 minutes of Lisa's social studies class and saw how respectfully Lisa listened and responded to her students' opinions on current events. This aspect of classroom instruction came very naturally to her and was not new.

Then, as math class began, Lisa brought out a set of base-ten blocks she had found abandoned deep in the school's storage closet. "We'll be working with just the special-needs kids today," she told Deborah. "I think I'm going to teach multiplication, but I don't know how."

But before Lisa had had an opportunity to introduce the lesson, her students, who had never used the blocks before, began to build towers with them. Lisa looked at Deborah, smiled, and shrugged. Her students were determined not to let anything interfere until their towers were built, so the adults just sat back and watched. When Melinda was satisfied her tower was finished, Lisa asked her how much, in terms of the values of the blocks, it was worth. As the other students finished with theirs, they came around to check out Melinda's, suggesting their own counting schemes. The rest of the lesson involved sharing counting schemes for all the towers the students had built.

"I had to go with where they were," Lisa explained in her conference with Deborah after class. "I saw they had to become familiar with the materials. But what mathematics did they do?"

Deborah proposed they analyze the various counting strategies, and then they discussed how these schemes illustrated properties of the number system—commutativity and distributivity—as well as providing practice in addition and multiplication of four-digit numbers. By the end of the conversation, Lisa realized that she and her students had been engaged in mathematics after all, and she had a few ideas about how to continue the following day.

Within just a few weeks of that initial visit and discussion, Lisa began to feel growing confidence and excitement. She was thirsty to learn, and able to articulate what she was learning. After a workshop for participating elementary teachers early in October, she described her changing conception of the nature of mathematics:

> For me the real revelation today was to discover that math is really like every other subject. I never realized that math is *learning* how to ask questions and how to think about the questions you ask. I knew this about literature, political science, biology, etc. Why it never occurred to me about math eludes me.

She then reflected parenthetically on just why mathematics had occupied its own special category for her: "Too much trauma associated with math and learning math? Maybe traumas of the past explain why I flipped out this summer!"

LEARNING MATHEMATICS

This change in Lisa's conception of mathematics allowed her to look at her curriculum in a new way. As she and her students reviewed whole-number operations that fall, she became intrigued by the patterns and relationships she began to see, and shared both her excitement and her questions with her students. Soon they, too, were finding patterns and exploring why they worked.

Though she was pleased with the changes that were taking place in her mathematics class, Lisa was aware that they were not enough. "I don't know enough math to teach effectively," she explained. "It is imperative to keep learning more math to be able to teach it well."

Lisa took on her commitment to learn mathematics with great seriousness. Each week she prepared for Deborah's visit by compiling a list of questions, both pedagogical and mathematical. In response to her interest, Deborah brought problems for her to work on at home. At the end of January, Lisa began the mathematics course described in Chapter 4.

Aware of the centrality of fractions in the fifth- and sixth-grade curriculum and of the difficulties children have with them, Lisa decided to make fractions the focus of her study that year. Specifically, she set as her goal understanding the contexts in which multiplication and/or division of fractions is called for. Lisa's journal account of her own process of working on decimal problems (taken from Harel et al., 1989, and Owens, 1988) illustrates her tenacity, as well as the rigorous expectations she had for herself:

1. *On a highway, a four-wheel-drive car can go 7.5 km on each liter of gasoline. How many kilometers can the car be expected to go on 1.3 liters?*

   $1.3 \times 7.5 = 9.75$ km. At first I was going to divide, but then realized that in order to find 1.3 of something you need to multiply. I don't really understand why this is true and didn't (don't) understand why certain problems imply multiplication with fractions either. The only confidence I have that this answer is probably right comes from knowing that [the car] can go further on 1.3 liters than on 1 liter, so the answer has to be greater than 7.5 by $\frac{3}{10}$ [of 7.5].

2. *Tom spent $900.00 for 0.75 kg of platinum. What would be the price of a 1 kg bar of platinum?*

The way I did this one was to convert to fractions. ¾ of x = 900 so ¼ [of x] = 300. Add 300 to 900 and you get 1,200. 1 kg would cost ¼ more and you already know what ¾ [is]. I wouldn't know by just looking at the decimal problem which operation to use without first going through what I just described.

While she solved each problem successfully, Lisa was still dissatisfied. She was looking for the generalizations that characterize multiplication and division, but as yet they eluded her.

On the decimal homework I had more trouble than I thought I would and my problems are the same as the ones I have with fractions. I couldn't see immediately . . . what operation to use for problem 1 and had to think it through as a fraction problem. I found myself doing that a couple of times, in fact, and always on the problems where the operation of choice is multiplication. This shows me the conceptual connections between decimals and fractions and also that I still don't understand what multiplication is. This really bugs me.

Over the next week, Lisa continued to struggle with the same issues:

1. *Each package of typing paper weighs 0.55 kg. Adam used 0.35 of a package for his research paper. How many kilograms of paper did he use?*
    .55 × .35 = .1925—multiply. You want to know what a little more than ⅓ of ½ is. For formulistic reasons that I'll probably *never* understand, to find a fraction of something, multiply. The answer should be a little more than ⅙. .1925 is almost ⅕ so that makes sense to me.
2. *Marissa bought 0.46 of a pound of wheat flour for which she paid $0.83. How many pounds of flour could she buy for $1.00?*
    I can't really do anything with this problem. It's kind of abstract. Marissa paid almost a dollar to get almost ½ lb of flour, so if she paid a whole dollar she would probably get close to a ½ lb or a little more. Instinct says divide something into something else so you can get a price-per-pound figure, but I don't immediately see what. The voice of memory says .46/.83 as x/1.00.
    .83x = .46
    x = .5542168, which according to my initial, "common-sense" reasonings is a plausible answer.

Again, even though Lisa was able to solve each problem correctly, she was after a deeper level of understanding. What is it in the structure of each of the problems that determines which operation to use?

"More Decimals" brought up more of the same previous issues: namely, that I don't understand multiplication and division of fractions or decimals. Problem 1 was easy. On problem 2, I began by thinking that I couldn't solve it because I couldn't get my mind around it. I knew it was a division problem but couldn't explain why and therefore couldn't tell what to divide into what. Then I went to simple algebra, because I knew I could solve the problem that way, from memory. Whenever I go to algebra it is a cop-out. It means that I can't derive what the problem is about at its core or what to do about it just based on "logic" and the information given. I had this same series of traumas with all these problems, as I explained in the homework protocols.

What interests me about this is how dissatisfied I feel . . . when I can't derive an answer "logically." I don't care how valuable math instinct (unconscious) is, it doesn't help me feel that I *understand*, in any kind of profound way, what's going on in a math problem. Knowing what to do isn't any good. I need to know why.

Lisa was doing more than simply solving problems. She was searching for patterns that would allow her to construct the general principles that govern the operations. She knew that understanding had to come from within. She could listen endlessly to other people's explanations and read innumerable articles about fractions, but she had to construct the ideas. Without that, all of the rest was mere empty verbiage:

I am still struggling with multiplication of fractions. Why is ¼ of 40 a multiplication problem? From the Driscoll article [1983] I got a new label for this idea—fraction as "operator," but I don't know what this means.

As her instructor, Deborah wondered why Lisa was so unhappy with herself when she was clearly doing so much good mathematical thinking. She could solve all the problems and verify each solution with sound reasoning. Was her self-condemnation the result of habit? All her life Lisa had thought of herself as a failure in mathematics. Was she unable to give up that aspect of her identity?

By the end of the course, it became clear that this was not the explanation. Lisa was looking for a kind of mathematical thinking in herself that she could recognize and respect once she thought she had achieved it.

For example, an activity sheet on area and perimeter included a problem that required drawing a trapezoid whose area was 36 square units. Lisa worked on this problem by drawing a series of trapezoids, each with an altitude of 4 units. Her first guess had bases of 15 and 7 units; her second had bases of 14 and 6 units; and her third had bases of 13 and 5 units. (See Figure 7.1.) Not only had she solved the problem from the activity sheet, she had also discovered

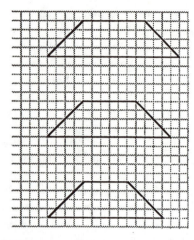

bases-15 and 7 units
height-4 units
area-44 square units

bases-14 and 6 units
height-4 units
area-40 square units

bases-13 and 5 units
height-4 units
area-36 square units

FIGURE 7.1. Searching for a trapezoid whose area is 36 square units

a pattern: "Every time you eliminate one unit from the bases, you lose four square units of area if the altitude is 4."

Curious about this discovery, Lisa now wanted to check trapezoids with other altitudes. When she drew one with bases 7 and 1, altitude 3, and another with bases 8 and 2, altitude 3, she saw that the area of the second had increased by 3 square units. (See Figure 7.2.) She described her process in her journal:

> I got into problem 15 and fooled around with trapezoids for a while. I noticed a pattern which was a little more complicated than the things I generally notice. I was also able to verbalize it more clearly and precisely than usual. Then I actually had enough curiosity (excitement? investment?) in this discovery to try to generalize from it, which may be a first for me. I also had a dim sense or suspicion of what it might be all about. I patted myself on the figurative back and said, Hey, you're improving. Your thinking is becoming a little more sophisticated, more precise. You're finding ways to test ideas. (You're actually *having* math ideas.) I had been working a pretty long time. I was perfectly satisfied with what I had "accomplished" because it was more than I expected to.

Lisa then described how she had gone for a walk and, when she returned, had sat down to continue her journal entry. And as she did so, she suddenly realized she could draw a diagram to explain the pattern she had discovered. (See Lisa's diagram in Figure 7.3.)

> I all of a sudden could explain the pattern I saw, and I'm pretty sure I'm right.

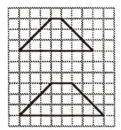

bases-7 and 1 units
height-3 units
area-12 square units

bases-8 and 2 units
height-3 units
area-15 square units

FIGURE 7.2. Testing a conjecture: if the altitude is n units, increasing the bases by one unit will increase the area by n square units

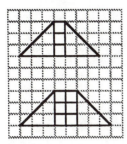

bases-7 and 1 units
height-3 units
area-12 square units

bases-8 and 2 units
height-3 units
area-15 square units

FIGURE 7.3. Lisa's demonstration convinced her of the truth of her conjecture

This whole process just described illustrates what it means to know a minuscule portion of mathematics, but it is certainly a starting point and I'm thrilled. Previous to this I would describe what I had learned since SummerMath as knowledge *about* mathematics, but I think that finally I'm beginning to know math and to understand, in a way that's meaningful to me personally, what that knowledge is. Feels good.

Now, after one year, Lisa was able to recognize that she had met the goal she had set for herself in her summer institute application: to acquire some confidence in her ability to think mathematically. Early in the year she had seen mathematics as a search for patterns and an exploration of the relationships among them. As she began to discover patterns she wished to explore, she could now generalize and show why those patterns must hold.

TEACHING MATHEMATICS

As Lisa worked on her own mathematical understanding, what was the impact on her teaching? Let us go back to December to see what was happening in her classroom.

Episodes from a Unit on Fractions

Near the end of her unit on whole numbers, Lisa confessed to Deborah that she was scared of starting fractions. "I don't know where to begin. We don't have any manipulatives, so I can't just hand them out and watch the kids. Besides, I don't even understand fractions myself!"

In response, Deborah gave Lisa a bare outline of where to start and what to cover—what fractions are, equivalent fractions, and operations—and handed her a set of nonroutine fraction problems to explore on her own. While working on these, Lisa accepted the challenge of devising her own problem sets for her students. To that end, she had to analyze fraction operations and think up physical situations that modeled them.

One day in early December, Deborah entered Lisa's school and, heading for her classroom, found several students sprawled in the hallways outside her door. As Deborah entered the room, she saw the rest of the class either lying on the floor or bent over the table, working on large sheets of oak tag.

"Hi," said Lisa. "We got started early today. They're making their fraction bars."

Deborah squatted down to watch Sonia, who had divided her sheet of oak tag into horizontal strips, each three inches wide. The top one was labeled "one whole"; the second was divided into two equal parts, each labeled "½"; the third was divided into three equal parts, and so on. Sonia was now working on the seventh strip.

"Shoot," she said. "It's still not coming out evenly."

Indeed, Deborah saw that Sonia had marked off six equal segments, but that the last one on the end was much larger than the others.

"We didn't have commercially made manipulatives for fractions," Lisa was to explain later,

so we made fraction bars because [the students] needed some kind of physical representation. They'd measure it out and decide what one third was, one half, fourths, fifths, sixths. So by the time we got to doing problems, they knew all about equivalent fractions and all about like denominators—from creating them.

There was a lot of discussion. You had an 18-inch whole; how do you divide that into sevenths? It won't go in evenly. And so it took them a while to figure out that you take that piece at the end and divide that into sevenths, and if you still have a little piece at the end you divide that into sevenths and then you have to figure out how to add those pieces to make each individual [seventh of the 18 inches]. It took them a while to figure out how to do it. It was very interesting.

Lisa understood that the purpose of making the fraction bars was not just to have them to use later, that making them was a valuable learning exercise in itself. She committed a full two weeks to the process to make sure the students learned how to use a ruler, understood that the pieces in each strip must be equal, and fully addressed any other questions that arose—how to make sevenths, for example.

> Then we cut all the pieces out [and laminated them] and put them in some kind of order. We talked about what the order was and what the rule was. Some kids put them all in wholes and some kids went from tiniest to bigger and some kids went with pieces that would fit evenly in other pieces [for example, halves, fourths, eighths; then thirds, sixths, twelfths, etc.]. So all of them were different.

By listening to her students as they worked at these activities, Lisa had the opportunity to find out what they understood so that she could plan the next phase of study. As she structured each activity of the unit, she considered what her students knew, what they didn't yet understand, and what they needed eventually to learn.

> The first problem set was pretty easy. I made the problems the weekend before I'd use them. . . . I'd watched [the students] with the fraction bars and I knew how much they knew already, so I didn't have to do anything with $\frac{1}{2} + \frac{1}{2}$. I could do really hard stuff—multiple-step problems. I knew that from watching them and interacting with them. I listened to their arguments about what patterns existed.

On a later visit, Deborah found Lisa's students working from another of her problem sets. Some of the students had gathered in pairs, some in groups of three or four; some worked alone first and then joined others to discuss the problems. Deborah sat down on the floor next to André, who had a pile of fraction pieces next to him.

"See, the problems are all about what kids in the class did over vacation," André told her. "This one's about Kwan Minh. *Kwan Minh and her family went skiing for December vacation. They spent $\frac{3}{5}$ of their time on the slopes and $\frac{2}{6}$ of their time riding to the top of the mountain. What fraction of their time was spent in the lodge sipping hot chocolate?*"

Knowing that Deborah would ask him what he was doing to solve the problem, André started explaining even before she had a chance to ask. "So this is what I have to do. This whole strip stands for the whole time of the ski trip. Then I've got $\frac{2}{6}$ and $\frac{3}{5}$." André sifted through his fraction pieces, selected the appropriate ones, and lined them up on top of the whole strip. "I've got to figure out what fits into that last section. The $\frac{1}{9}$ is too big," he said, testing it out. "So is the $\frac{1}{10}$, and so is the $\frac{1}{12}$. The $\frac{1}{15}$ looks pretty good."

Deborah looked down and saw that the $\frac{1}{15}$ did look pretty good. But André's fraction pieces hadn't been measured and cut accurately. Everything was just a bit off, and that $\frac{1}{15}$ hung over the edge of the strip underneath just a little.

But again, André was ahead of Deborah's questions. "It's just a little bit off," he said, "so now I have to figure out if it's because my pieces are a little off or if it's really not exactly $\frac{1}{15}$."

André wrote down: $\frac{2}{6} + \frac{3}{5} + \ ? = 1$.

"So," he explained. "I can't add them like that. I have to change the numbers. See, I can make the $\frac{2}{6}$ into $\frac{4}{12}$, but that doesn't help, and $\frac{8}{24}$ doesn't help either. Oh, wait! $\frac{2}{6}$ is the same as $\frac{1}{3}$. So, let's see, $\frac{1}{9}$s? $\frac{1}{12}$s? $\frac{1}{15}$s? Yeah! I can use $\frac{1}{15}$s—$\frac{2}{6}$ is the same as $\frac{1}{3}$, and that's the same as $\frac{5}{15}$. And $\frac{3}{5}$ is the same as $\frac{9}{15}$." And he wrote down on his paper: $\frac{5}{15} + \frac{9}{15} + \ ? = 1$.

"See, look. This comes out to $\frac{14}{15}$," he said, pointing to $\frac{5}{15} + \frac{9}{15}$. "So this has to be $\frac{1}{15}$. I was right; it's $\frac{1}{15}$."

What might have been a liability of these home-made manipulatives had, in this case, actually become an asset. Having made the fraction bars himself, André was fully aware of their inaccuracy. While they provided a representation that allowed him to picture the terms of the problem, they did not give him an answer. He knew that "close enough" was not satisfactory, and the inaccuracy of the manipulatives forced him to think through the mathematical structures in order to conclude finally that, yes, Kwan Minh had spent $\frac{1}{15}$ of her skiing vacation sipping hot chocolate.

Lisa described the classroom interactions that had preceded André's solution to the problem:

> At the beginning [the problems] all had like denominators, addition and subtraction. Toward the end of that problem set I threw in a couple of easy unlike denominators. Then we tried to derive what the rules were for addition of fractions for like denominators and addition of fractions for unlike denominators. I realized they couldn't really verbalize what the rules were for addition of fractions for unlike denominators, so we needed some more problems.

In each problem set, Lisa always included a few problems that addressed the next topic to be covered. In this way she would assess what her students already understood and where their difficulties lay.

> What I do the next day is based on what I observe and what I hear, how well the kids are able to verbalize what they're doing, how interested they seem. . . . What I'd do is always throw these teasers in and then they'd

realize they needed more information and of course I wouldn't tell them. So then I'd be compelled to design more problems. . . .

[In response to the teasers, they'd say,] I can't do this! And I'd go, Well, why can't you? Well, because you can't add those things. Well, show me with the fraction bars. They'd say, Well, it's easy, you just do this and this. And I'd say, Write an equation that shows what you just did. And that's the hard part. And I really pushed that.

Some weeks later, the class was working on multiplication. They were all sitting in a semicircle, some in chairs and some on the floor, looking at the blackboard. "How can that be?" asked Ivan. "You just said that was a half. Now you're saying it's a fourth."

Larry was up at the board pointing to his diagram. "Look, the problem says Lelieta's sister left her half the cake. So, see, here's the whole cake and here's the half that her sister left. Now, Lelieta didn't eat that whole half. She ate only half of it." Larry was marking off the amount Lelieta had eaten, pressing the chalk down as hard as he could. "So, it's a fourth of the cake. See? It's a half of what her sister left, but it's a fourth of the cake."

"What do you think, Ivan?" Lisa asked.

"Yeah, I guess. It's like, if I watch half an hour of TV, that doesn't mean that I watched half the night. Just because it's half an hour doesn't mean it's half of everything."

"Can anybody paraphrase what Ivan and Larry are saying?"

The discussion continued, going through other problems the students had worked on, examining how a given amount might be represented by different fractions. By the end of the period, Lisa felt that her students' understanding of that idea was solid enough for the class to be given another question to wrestle with.

"You know, we said earlier that all these problems were multiplication problems. And you all know how to solve them. Well, here's something else to think about. When we worked on multiplication with whole numbers, we understood it to be repeated addition. Right?"

All the students looked at her and nodded.

"But look at this. $\frac{1}{2} \times \frac{1}{4} = \frac{1}{8}$. I can't see any way in which anything is repeated. So if it's no longer repeated addition, what is multiplication?"

On that note, the math period ended and the students went off to music class.

When Deborah returned a week later, Lisa told her what had happened the following day. "Remember the question I posed? Well, the next day I asked it again at the beginning of the period, and we talked about it for a long time. By the end of the class, everyone was completely confused. So then they turned to me and said, 'Well, what's the answer?' And I said, 'I don't know. I'm as confused as you are.' So then they got really mad and said I shouldn't ask them questions that I couldn't answer. And I said, 'Why not? It's an important question. I'm just learning this stuff now, too, so we're going to figure it out together.' It was great."

By the end of the unit, Lisa was quite pleased with the learning that had occurred in her class—both her students' and her own.

> At the end of the fractions unit they were just beat. We had learned a lot about [fractions]. They were still pretty confused about why "of" means multiply, which is something I don't understand myself so I can't even help them. So these things are still issues. But they were thrilled with themselves when we got to decimals because within a week they understood almost all there was to know about them. They said to me, I can't believe this is so easy. And I said, that's because you were doing so much of your decimal work when you were doing fractions.

Establishing a Mathematics Community

At the beginning of the year, Lisa had worked to establish the idea that she and her students constituted a community engaged in a common process of inquiry. Yet, though her students had gotten used to collaborating in their other subjects, they did not automatically transfer these modes of interaction to their mathematics lessons. Prior classroom experience as well as deeply entrenched societal attitudes convey messages about what mathematics is and who can do it that often undermine the possibility of genuine community.

Among those messages, Lisa's students carried into her classroom the fundamental principle of traditional mathematics instruction: correct answers are the goal and confusion is shameful. Those students who had always been successful at rote computation were especially reluctant to change. They were used to being recognized as top students without making much of an effort and had no interest in taking on a challenge that might expose them as less than the best. Nor were the weaker students interested in engaging in a process that might render their coping mechanisms ineffective.

However, as Lisa persisted in asking probing questions, as she responded positively to the ideas her students offered, and as she consciously worked to create a safe environment, she soon persuaded her students that she expected every one of them to learn and to contribute.

"If the expectation that everyone is capable of understanding and solving the problem is established at the outset," she explained later, as she reviewed her year of instruction,

> [and] if kids are encouraged to debate and argue, to be wrong but interesting because it may be useful to the rest of the group as they struggle to-

gether with a problem (like why do you invert and multiply when dividing fractions?), they will persevere.

The inevitability of confusion, as well as the discomfort, became an acknowledged part of the ethos of the group:

The old bugaboo [for me] remains division of fractions, which sometimes I understand, and at other times don't. This drives me crazy, since I never know when I'll get up to the blackboard with a kid to help with clarifying questions and then have to deal with internal white-out. The kids . . . call this process of disunderstanding "entering the Math Zone." When some one enters the Math Zone we all wax deeply sympathetic and hum in unison the Twilight Zone theme. This doesn't help any of us deal with the confusion engendered, but knowing that it is a recognizable place for most of us does lend the courage of camaraderie to what for me as a kid had always been a devastating and debilitating experience.

This recognition made possible the kind of learning that leads to the construction of powerful mathematical ideas.

They understand that ½ can be ¼ at the same time. [How did they learn that?] . . . They were motivated and excited by the ideas involved, and they weren't quite as scared as usual to be wrong because they had seen everyone, including myself, wrong. Because it was cool to be wrong as long as you're interesting!

Yet, even as this new atmosphere established itself in the classroom, Lisa became aware that some negative social norms continued to operate. Most dramatically, Lisa realized several months into the year, in this class of 8 girls and 15 boys, the girls had no voice. Gender patterns into which the children had been socialized in previous mathematics classrooms, patterns reinforced by societal attitudes, had become more obvious (Belenky, Clinchy, Goldberger, & Tarule, 1986; Chipman, Brush, & Wilson, 1985; Damarin, 1990; Fennema & Leder, 1990; Gilligan, 1982). Lisa had established a classroom dynamic of active and sometimes excited exchange of mathematical ideas, but the boys dominated discussion, while the girls, reticent about participating, were marginalized.

The girls are very reluctant to compete with the boys. They are so afraid of being considered wrong and being made fun of. . . . It's the worst in math. They felt the worst about themselves mathematically.

Lisa saw a way to address the problem when, during a discussion with a mother who was concerned that her daughter had already given up on mathematics, Lisa suggested that the child stay after class on Wednesdays with a friend and do mathematics. But then, on second thought, why not encourage all eight girls to come? So the "girls' math club" was formed.

> We started out just doing what we were doing in class after school, and then we ended up getting there well before the rest of the class, and we ended up doing just lots of fun stuff. . . . [There was always a] high level of noise and excitement. [They] would often solve problems while jumping up and down and singing. [I] brought the principal in a couple of times to witness this and to see long division and four-place multiplication in Xmanian. . . . I would have to *throw* them out at the end of an hour and a half. This is after a full day of school.

The formation of the girls' math club significantly altered the classroom dynamic.

> What the girls mastered . . . changed . . . their voice in the class. The voice of the classroom had been extremely masculine and what they did was to change the quality of that voice.

Rather than being intimidated into silence, the girls "came to realize the kinds of mistakes [they now saw] the boys were making . . . they'd been making [them] all along, but [the girls] hadn't been catching them."

The proactive intervention of the girls' math club was, in this case, necessary to counter the self-defeating attitudes that these girls had already developed about themselves and their mathematics ability.

Students' Responses to Mathematics Class

As a combination fifth- and sixth-grade teacher, Lisa taught some students two years in a row—both in the year prior to her participation in a SummerMath for Teachers institute and in her first year of active involvement. This provided an opportunity to explore these students' perspectives on how Lisa's mathematics instruction had changed. As it turned out, four boys—Ivan, Adam, Jacob, and Alex—and two girls—Lelieta and Ogechi—volunteered to be interviewed. Ivan and Adam had had Lisa for both fifth and sixth grades. Although the interviews were conducted in pairs, for the sake of readability the comments of all six students have been consolidated here into one discussion.

> *Interviewer:* Ivan and Adam, did the way Ms. Yaffee taught math change from one year to the next?

Ivan: Yeah. Last year we just kept working in a book on our own. She didn't assign us pages. We just kept going on our own.

Adam: This year we learned a lot more. We learned why the problems worked, why the method worked. Last year we just did the stuff. If there was something new, like how to multiply fractions, Ms. Yaffee just told us, or the book did. It wouldn't ever say why.

Ivan: With math books we'd just learn the same basic things over and over again, but maybe the numbers would be a little bigger or harder. What we're learning now is a lot more complicated.

Adam: With the textbook, we'd just go through the book and we didn't really go over what we needed to know. Ms. Yaffee this year would give us a worksheet she made up, or we'd discuss it, and then she'd see how we'd do and if we looked like we needed . . . to know [more] then we'd start a section on that.

Interviewer: What about the rest of you. Was math class different for you than in past years?

Lelieta: It's a lot funner because it's not boring.

Ogechi: In previous years we really didn't ask questions. If you didn't know how to do problems in the book, [the teacher would] show you again and again until you knew how to do it and you could do it. You were supposed to memorize, This is right, this is wrong; do this, don't do that. You're supposed to do it this way.

Jacob: [Now] you have a problem and you do it one way, and [Ms. Yaffee] wants to know why you do it that way, why you chose it.

Ogechi: Sometimes if we didn't get it, we'd go over to other students who got it and if we didn't get it from them then we'd ask Ms. Yaffee and she'd try to make us understand by putting it in a different form. She didn't tell us why it works. Maybe she'd give you an example of some sort and from an example you might figure out why it works.

Lelieta: [When we could answer our own questions] we were excited about learning this new thing. It's a whole 'nother world.

Interviewer: I understand that you usually worked in groups. What was that like?

Adam: It's good to work in groups because when you get older you'll have to work in groups to get something done. How to cooperate and let other people talk.

Alex: By working in groups we also hear other people's point of view on the things, and hear things that we couldn't think of by working alone.

Ogechi: I like this year's math a lot better 'cause one of the main things is that last year there weren't any discussions.

Interviewer: You mean discussions in your small groups?

Ogechi: Yeah. But also with the whole class.

Alex: This year it gave us a chance to express our feelings about some of the problems and if we didn't understand we could get it cleared up. But not the answer. Other people's opinions.

Interviewer: What were the discussions like?

Lelieta: In group discussion, Ms. Yaffee asks the questions and sometimes the kids ask each other questions: How can you prove this works? Can I see how you prove this? Clarify how you prove this because I don't understand. How did you come up with this?

Jacob: We had a lot of conflicts.

Interviewer: How were conflicts resolved when students had different points of view?

Ogechi: [To decide if something's right or wrong] we'll have arguments between people. Then we'll find out . . . if it's a reasonable answer. [Sometimes it turns out] both of the persons are right.

Alex: Another way to see if a method is wrong is applying the method on whatever problems we're doing. Like, we have word problems every week. If someone suggests a method then we use that method and if we can apply it to the problems that we're doing and it doesn't work with what we're doing [then it's wrong].

Interviewer: You've talked a lot about how the class was different and how you understand things better. What about the math itself? Was what you studied any different?

Jacob: What we're doing now, making banquet tables out of squares [an activity Lisa had taken from Shroyer & Fitzgerald, 1986]. That's different. . . . Here we can see how with the same area you can make the perimeter bigger by changing the shape of the rectangle. In the book, it just says area is this and perimeter is this. But she does things like find the biggest perimeter and the smallest perimeter using 12 squares.

Alex: [And for fractions] we'd have to figure out a rule. We'd do these problems over and over again until we could see a rule and why it works. . . . We'd get math sheets with three pages and they'd be all different word problems and then we'd talk about what we did and then we'd talk about why something works. We'd just have to figure it out by doing the problems.

Jacob: Once we got through fractions and stuff, all of us understood numbers well enough that we could do the rest of the math things really easily. . . . Fractions is about understanding numbers and once we understood numbers we could do everything else really easily.

Alex: Another thing that's good this year is that Ms. Yaffee doesn't already know [a lot of the math]. She's just like one of the students.

She just asks questions. She doesn't know the answers. She doesn't
really know so it's not like she's superior to us.

Lelieta: When [Ms. Yaffee] was growing up she didn't know why you
subtract when you do long division. People just told her to do it.

Ivan: She knows more now, though. And [next year] it will be not as
good because then she'll know the answers so it won't be as good
because there won't be so many arguments.

Ogechi: I think both us and her have learned a lot this year. . . . And I
think . . . the class won't be as good next year because we're feeling
that the teacher isn't that much more superior. She knew, but she
didn't understand.

Adam: I hope next year in seventh grade math will be the same as Ms.
Yaffee did it.

LISA'S REFLECTIONS

The changes in Lisa's mathematics instruction grew primarily out of her
developing understanding of the discipline itself. In contrast to Jill Lester (Chapter 5) and Sherry Sajdak (Chapter 6), Lisa began with a well-developed understanding of constructivism that formed the basis of her instruction in other subjects. Jill started the school year with the idea that she wanted to use some new
strategies in her classes, so her main task became to understand student learning
in mathematics as a developmental process. And while Sherry struggled with
contradictions between equally compelling but very different conceptions of
teaching, unless she got some new classroom strategies working, the bigger
questions would never get sorted out. But Lisa began with a strong sense of
what her mathematics class should be like, and she already knew how to pay
close attention to her students' learning. What she needed most was help in
exploring mathematical issues related to the upper elementary curriculum. She
also needed to develop confidence in her own power as a mathematical thinker.

[Last summer] I was panicked and really upset because I knew I couldn't
go back to doing what I had done. [But the institute] didn't give me the
real basic practical stuff I needed. Then Deborah started coming on
Wednesdays. I could say to her, How do you teach this? How do you
teach that? What would you do about this? What would you do about
that? And she had really concrete suggestions and materials. When I'd be
leading my first tentative discussions and I didn't know what to say anymore, I'd just be standing there, but I could throw it to her and she would
sort of lead me out of the mire. . . . So that's what changed it. Follow-up.
What I loved about it was, there was no agenda. It was math therapy.

I could go in and go, Oh my God, what do I do about this? I'm really confused about this. Someone asked me this question and I can't answer it. And whatever my agenda was, was the agenda. And that was great.

Then she would say to me, Here are some problems. Why don't you look over them in your spare time? And I would. And I'd come back and say, I couldn't understand this, and we'd discuss it. And I could say things to her like, Well, what is the theory about this? It was like having a journal, but a journal with clarification, you know.

When Lisa was interviewed at the end of the year, she gave her overall sense of how things were going: "I do feel like it's working. I do, and I feel that next year I'll know more about what I'm doing." Then she described how she had changed from September to June.

I'm a lot less nervous. I'm a little more confident, not a whole lot more confident. I'm only beginning to think mathematically now. I'm very shaky at it. For me [confidence as a mathematician and confidence as a math teacher] are very closely related.

When her interviewer told Lisa that her students felt she would never be as good a math teacher again because this was the year that she was learning along with her students, she responded:

I understand perfectly well that I can teach math well without understanding math. . . . I learned to use my not knowing to work for me. . . . But I don't like it. I feel like I could ask much better questions and write much better problems if I had a clearer understanding of what's involved. [My not understanding] limits a lot of what I can do.

"Besides," she said, "I have a lot more math to learn." When Lisa applied for another mathematics course offered by the program the following year, she listed some of the topics she now wanted to study:

I need to understand ratios, and how they differ conceptually from fractions. I need to be able to create a meaningful context for exponents, in addition to a better concept of their nature (including negative exponents and fractional ones). I need a better understanding of what negative numbers are. An exploration of how to define variables and how to use them would be useful, as would some context for graphing functions. I would love some theoretical background on what mathematics *is*, according to "them what knows." I've heard terms like "function," "relations," and so forth. What do they mean? My background in geometry is abysmal. What

is it? What good is it? Why is space considered a mathematical entity?
. . . What are trig. and calculus and why were they invented? Why is
something raised to the 0 power one instead of zero?

"I'll tell you, Deborah," she wrote to the instructor, "you've got your work
cut out for you."

WHAT DID LISA KNOW?

A point made repeatedly in this book is that, in order to be able to teach
mathematics for "understanding" in the sense argued for here, teachers must
know more mathematics than they, or most of their fellow citizens, currently do.
Yet, Lisa's story poses a paradox. On the one hand, Lisa candidly admitted that
she did not fully understand the mathematics she taught and frequently set problems for her students that she herself could not solve. On the other hand, she
had become an unusually effective mathematics teacher. Her students could describe the ways in which their own mathematical understanding had improved
and could even identify what Lisa had done to bring that understanding about.
In fact, they believed that Lisa was a more effective teacher precisely because
she didn't know the mathematics beforehand.

Resolution of this paradox lies in an appreciation of what Lisa did, in fact,
bring to her mathematics instruction.

First, and perhaps most important, Lisa began with an appreciation of what
it means to really learn something—"how to ask questions and how to think
about the questions you ask." Once she had confirmed subjectively, in her own
struggles in the institute and in her work with Deborah, the relevance to mathematics of her theory of learning, she was able to overcome the paralysis that had
affected her mathematics instruction. She decided that, just as she had done in
other subject areas, so in her mathematics lessons she would aim to help her
students learn to pose their own questions and would give them ways to think
them through. She decided, too, to eschew pretense and use her own ignorance
in order to model this process for them. By inviting her students to help her with
her own questions, she could show them how to listen to others' ideas and demonstrate to them the value of collective inquiry.

But, in addition to her constructivist orientation, it is important to think
about the mathematics Lisa knew. Here an analogy may be helpful: the teacher
as guide. Leading her students into terrain unfamiliar to them and only imper-
fectly understood by her, she did have a map that identified at least some of this
region's major topographical features. In order to begin to fill in the map—of
fractions, for example—Lisa scouted some of the main concepts by herself, first
with Deborah's help and then in the spring course. Through this work she was

able to survey the overall terrain. She understood that among the major areas she needed to explore with her students were equivalent fractions, addition and subtraction with like denominators, addition and subtraction with unlike denominators, multiplication, and division. She also identified some of the big ideas that defined the geological substructure of the region: that the same object can represent fractions of different values, depending on the reference whole; that the referent remains constant when one is adding or subtracting, but changes when one is multiplying or dividing; that some generalizations about whole-number operations (e.g., a product is almost always larger than its factors; a quotient is almost always smaller than its dividend) no longer hold; that the very meanings of multiplication and division must be extended beyond the meanings derived from whole-number operations.

All this enabled Lisa to select a route—to write word-problem sets—that would lead her students to the high ground from which they could view the surrounding terrain. She understood, too, that in order to monitor her students' developing maps so as to determine which trails to take next and where they needed to explore further, she had to listen carefully to their ideas and challenge them to articulate those ideas clearly. Lisa had to distinguish valid arguments from invalid ones, lead her students through their misconceptions when they wandered from the trail, and even fill in swampy ground to make it traversable.

With these kinds of knowledge and understandings, Lisa could design lessons and conduct classes in ways that enabled her students to construct their own internal maps of fractions country. That did not mean, however, that she needed to have already charted for herself each trail her students would need to go down, each obstacle they would encounter. It did mean that she had to be prepared to acknowledge that she, too, had lost her way, that she had more exploring to do. "At times," Lisa said,

> I would be confused; I would make mistakes; and I would not know what I was doing. And the kids could clearly see that. I would become confused over a problem that I didn't really understand. . . . There was a division-of-decimals problem [*Tali weighs 61 pounds. If his weight is .80 of his best friend's, how much does Tali's friend weigh?*] that I had written that I didn't understand. . . . Alex, Jacob, and Lelieta understood it and I didn't, and they knew that. And I said to them, I can't understand this, so I'm just going to have to work on it myself.

Though Lisa was candid about the limits of her own understanding, her students did not lose respect for her. They recognized that she was being honest with them, and, by the example she set, they learned that it was not shameful to admit ignorance. Through her own efforts to understand the mathematical ideas, Lisa communicated the importance of what they were learning.

Lisa's students could see that she often did not understand the mathematics they were learning, but they also knew that they were not wandering aimlessly: the lessons were organized and focused. They trusted Lisa to guide them, and their awareness of the power of their own learning—when they looked back down the trail they could see the progress they had made—reinforced that trust.

Thus, the question is not whether Lisa understood all the mathematics she was teaching, but rather, What did Lisa know that, in spite of the gaps in her understanding, made her so effective?

This chapter has described Lisa's struggle to generalize the patterns she discovered in the multiplication and division of fractions problems she worked on. For example, to solve one of those problems (*Marissa bought 0.46 . . .*), she first estimated that the answer was close to or just over one half, and then calculated the result, .55. Yet she wrote, "I knew it was a division problem but I couldn't explain why. . . . [I want to know] what the problem is about at its core."

And though some of her students were able to solve the problem about Tali's friend's weight, Lisa insisted that she herself didn't understand the crux of the problem. But her confessions of ignorance need not be taken at face value. After all, *she* had written that problem in order to explore division of decimals, and she had solved it. But it was something else she was after—"What is it in these problems that makes them division problems?"—and, though she didn't yet understand that, it was she who had helped Alex, Jacob, and Lelieta to see further, at that moment, than she herself was able to see.

By posing that question to the class—"What makes these division problems?"—she did more for them than untold hours of mechanical practice at an algorithm could have accomplished. For in listening carefully to her students' ideas, she validated the power of their thinking. And by working at the edge of her own mathematical understanding, Lisa not only enlarged her students' opportunities for learning, but she provided herself the best possible setting for the kind of learning she needed to do and now knew she was capable of.

Part III
BEYOND THE MATHEMATICS CLASSROOM

8
Learning to Take Risks: Ginny Brown

I'm now going for a master's, but before it had never occurred to me that I could get an advanced degree. It was my experience in this program that helped me see that I could learn. . . . This victory gave me the confidence to move on.

Joyce Zippe
Woodland School
Southwick, MA

"It's cool to be wrong as long as you're interesting," Lisa Yaffee's students concluded midway through the year. In a classroom that valued critical analysis and the ability to think for oneself, they were acquiring a new standard by which to assess their learning. Students exposed to an instructional practice based in constructivist perspectives of learning come to realize that if they voice their ideas, those ideas will be heard and valued, and that even if their ideas don't lead to solutions to the problems they are working on, they might well spark ideas in their classmates. Their standard is no longer who can get the most right answers in the shortest time. Rather, these students come to value their own and their classmates' abilities to think and to question.

As teachers restructure their lessons around group problem solving and emphasize attention to the reasoning process, they note changes in their students that extend beyond enhanced mathematical understanding. They report that their students are more willing to take risks, are more self-reliant, in short, that they are in general more confident. And the respect that these children are acquiring for their own ideas about mathematics carries over to other aspects of their lives (Schifter & Simon, 1991).

It is not surprising that teachers who make these kinds of changes in their instructional practice experience analogous shifts in how they assess their own competence. Like their students, they, too, become more willing to take risks, and their confidence grows accordingly. In some, this is expressed as greater intellectual curiosity and increased interest in the profession. These teachers may take on more active roles in shaping curriculum and staff development programs, or may even choose to pursue graduate degrees. This chapter tells the story of Ginny Brown, whose third-grade class was visited in Chapter 1, and who, as a result of her work with SummerMath for Teachers, discovered in herself unsuspected strengths. The increased confidence this gave her led her to take on new challenges and to create for herself a richer professional life.

ENTERING THE PROGRAM

Ginny had always played it safe. By her own admission she hated change. For 18 years, she had taught third grade in the same school in a small rural town where she and her schoolteacher husband had lived most of their married lives.

Cathy first met Ginny over lunch one day when she was paying Jill Lester (Chapter 5) a follow-up visit. Ginny and Jill were colleagues with adjoining classrooms. After the initial introductions, Ginny volunteered shyly that she was beginning double-digit multiplication with her class, and that, though Jill had been encouraging her to use base-ten blocks, she was struggling with "how to show multiplication with the rods." After describing some appropriate activities to her, Cathy suggested that Ginny might want to just play around with the rods herself to find various ways to represent multiplication.

But Ginny seemed to have something other than multiplication on her mind. What she really wanted to talk about soon became apparent. "Jill's been encouraging me to enter SummerMath for Teachers next year, but, to be honest, I'm very nervous about having someone in my room once a week."

"Why?" Cathy asked.

"I don't feel comfortable being observed."

Seeking to reassure her, Cathy said, "Ginny, my role would not be to evaluate, not even to critique, unless you'd asked me for that. My role would be that of a resource to help you implement whatever it is that *you* wanted to do." Cathy hoped she had conveyed her willingness to accept Ginny's goals, whatever they might turn out to be. Ginny responded that she would continue to think about applying.

Several weeks later Ginny met Cathy in the hall and said that she and a colleague from the school had both applied for the next summer. However, her fear of having someone visit her classroom was very real. Later she was to comment, "I thought Jill was nuts. . . . She *wanted* someone in her room! *Not me!*"

At the start of the summer institute, Ginny alluded to this fear in her journal. "I'm scared to death of having Cathy come into my room and see me flop." And a few days later she wrote:

It is now 5 AM and I woke up with a panic attack. How am I ever going to be able to do this in my classroom? . . . Day in and day out for 180 days? . . . I guess with Cathy's and Jill's help "I shall overcome," and after bumbling my way through the first year of it, *maybe* it will become more of a natural thing. . . . But I am still dreading the weekly "observations" (in spite of Cathy's assurance that it is a "consultation" and not an "observation"!).

Ginny's lack of confidence extended beyond her teaching. On her application she had described herself as weak in mathematics and as having little to contribute to the program:

> To be honest, my own elementary-school experience was such that I never clearly understood the process of regrouping until I began using the manipulatives [as an adult]. . . . I am really not certain what I can contribute to the project. Perhaps, having been exposed to the results in [Jill's] classroom next to mine, I can contribute enthusiasm for the methods involved.

The first few problem-solving sessions drove Ginny's confidence even lower. In fact, when she was interviewed a few years later, she still remembered the anxiety she had felt the first day of the institute. "In high school algebra," she explained,

> I often felt lost, but I could memorize and learn rules even when I didn't understand. The first day of the institute we started with a very simple problem that had to do with planting trees. It seemed like everybody at my table knew what to do, and I felt awful. Formulas floated around in my head and all my old phobias came back.

A CHANGING CONCEPTION OF MATHEMATICS

Ginny's conception of mathematics had been simple: get the right answer, the right way. She saw mathematics as a store of algorithms and believed that good mathematicians knew which algorithms to use when. On the evening of the first day of the institute she wrote in her journal:

> I am not really sure what my theories or beliefs are concerning how people learn mathematics. I'm not even sure if I have such theories and beliefs. I guess there are many ways people learn math. The math I learned was learned by listening to teachers explain various mathematical concepts, demonstrating on a chalkboard. I never considered myself "good" at mathematics. . . . I had to work hard to grasp it. Some people seem to learn math a lot more easily than others. . . . I'm not convinced that everyone is as capable as everyone else when it comes to [doing this]. I'm sure I have a harder time with math than some.

As the institute progressed, Ginny's conception of mathematics, and her relation to it, began to change. After an activity that involved estimating the volume of an irregularly shaped room using a shoebox as a measure, she wrote:

This problem-solving session regarding volume caused me to realize that volume is not a cut-and-dried formula of length × width × height. Sometimes different ways of measuring need to be used for the most efficient use of estimating the existing space. The fact that there were different ways to interpret this problem did not take away from the value of learning about volume. In fact, I think it added to it.

Ginny was referring to the fact that several participants had understood the problem to be about how many shoeboxes would fit the space, while others had used a side of the box as a standard of length and had estimated the volume of the room in cubic units. A somewhat heated discussion had ensued over the correct interpretation of the problem. It had ended with the insight that adopting conventions by group consensus is part of the mathematical process, too. Ginny's thoughts returned to this discussion as she wrote in her journal that evening:

The belief that an answer is either right or wrong is so engrained in the majority of teachers. . . . It became obvious from today's session that this will be difficult to break. . . . Perhaps this is one of the pieces that this program is all about.

The Xmania activity provided Ginny with further insight into the role of convention in mathematics. "When dealing with math, many things are arbitrary and many things are set. I think that previously I believed not much in math was arbitrary." Xmania also taught her that "Discovering something for oneself is so much better than having someone tell a solution to a problem!" In her journal she pursued this theme:

The Xmanian problem brought up more questions than the original problem posed—so much more learning was going on than originally planned! Some unanswerable questions (at the time) were brought up—but not understanding something really seems to be equally important as understanding some things, as long as it perpetuates the thinking process.

Ginny's static view of mathematics as a set of truths, facts, and formulas, which some people know and most others do not, was giving way. Like Linda Sarage (Chapter 4), she was learning that mathematics is derived, or constructed—partly as a matter of logic, and partly as a matter of convention. Furthermore, she was learning that she, too, could really explore mathematical ideas; that she, too, could participate in the construction of mathematics. As

Ginny began to enjoy the challenge of trying to figure things out for herself, she came to recognize, as Linda had, that feeling puzzled could be seen as just a stage in the process of learning. Like Lisa and her students, Ginny could pause to laugh at her puzzlement and then continue struggling with a problem, knowing the satisfaction that would come when she finally found her own way to a solution.

A CHANGING CONCEPTION OF TEACHING

Ginny's changing conception of mathematics brought with it a newfound confidence in her ability to think mathematically. But it also undermined her view that the job of a teacher was to transmit understanding to her students. Though the realizations were painful, she reflected openly and honestly on what she now felt had been the inadequacies of her past teaching. In a synthesis paper at the end of the first week of the summer institute, Ginny wrote:

> The concept of volume, which we worked on last Tuesday, could not have been grasped satisfactorily if the teacher had stood in the front of the room and presented it to the class. The Xmanian problem really put across the importance of allowing children to construct a solution to a problem themselves. I'm afraid I had never given a lot of thought as to how children learned math, or even, actually, what was really meant by "learning" math. I assumed that learning the mechanics of the various mathematical operations was sufficient. For children to do this it seemed logical, after presenting the material to them, to provide them with many opportunities for practice. For them to fully understand the reasoning behind the operations didn't seem important to me. I hate to admit this, but I'm not sure *I* fully understood the reasoning behind all the operations. Since I have had the opportunity to be a learner for the past week, my conceptions concerning how people learn have drastically changed. It has become obvious to me that in my math classes many children were not actually learning, but were being "programmed" with information. Processes were being memorized but not understood.

During the next week, Ginny began to construct a new view of teaching that reflected these insights. "We have a contract with students," she wrote in her journal, "to continually monitor what they are doing and be sure understanding is there. It's okay to let the learner go on the wrong track for a while if it will best enhance learning." A few entries later her ideas had solidified further:

> I now firmly believe that for every math concept I want to teach, I will need to start from what the children already know and allow them to con-

struct in logical sequence the learning that they will need in order to grasp that concept. The constructing that the children do will evolve logically if the lesson sequence is well thought out. It must include, for all concepts, problem solving, the use of manipulatives, small-group work, recording sheets, verbalizing, whole-class "pulling together" and, often, a pre-group-discussion period; and perhaps the most important component of all—questioning on the part of the teacher which "leads" a child to form his/her own conclusions, follow a path that he/she determines is helpful in solving the problem, or reflect on what he/she has just done. Ideally, if I am able to follow this model, I will no longer be a "teacher," but rather a "guider of learning."

Yet, even as Ginny articulated her new vision of teaching as facilitation, she still believed that there was one right way to facilitate—one right question to ask, one right lesson to follow up with—and this idea gnawed at her confidence in her performance. "I really worry," she wrote in her journal,

> when I'm trying to do this in my classroom, will I be able to ask the right questions without giving information? . . . It worries me that I will have trouble designing lessons that incorporate all these . . . characteristics, and that I will not remember to do all the important things inherent in the actual implementation. Maybe I won't know how to question effectively.

Thus, though Ginny was able to reject many of her old beliefs about teaching, she still believed, as Sherry had, that someone somewhere knew exactly what she should be doing—and, whatever that was, she was afraid she wasn't doing it.

But as Ginny worried about her ability to realize her new vision of mathematics teaching, and as her panic over Cathy's impending visits to her classroom persisted, she began to recognize the underlying causes of her anxiety.

> I am so used to the "show me" type of teaching that I resort to asking for too much help when I'm doing anything. Maybe it's a confidence problem. I guess I'm really afraid to take risks. Sometimes it seems as though I have to figure out how to do something with nothing to jump off from or no base from which to start.

TAKING RISKS ALONG WITH HER STUDENTS

Ginny was nervous when Cathy visited her classroom for the first time. She had asked Cathy to teach that first lesson, but Cathy had declined, explaining that she wanted her work to complement Ginny's. In reality, Ginny didn't need another demonstration lesson; she had observed Jill often enough the year before. She simply needed to overcome her fear of having Cathy in her classroom.

Despite her anxiety, Ginny's lesson went well. She had designed a story problem that required her students to subtract with regrouping. The children worked in pairs, using base-ten blocks, then met together as a class to discuss their solutions. After the lesson, Cathy focused their discussion on the children's understanding in order to reassure Ginny that she was there to reflect with her, not evaluate her.

For the next few weeks, Cathy and Ginny alternated taking the lead teaching role, presenting problems and facilitating discussion. The children were used to explaining and debating their solution strategies since they had been Jill's second graders the year before, and so they reacted to the lessons with enthusiasm. As the weeks progressed, Ginny began to relax and enjoy the way the children did mathematics.

"It's the response of my class," she explained one day when asked about her growing confidence.

> The kids get so excited about their solutions, and I feel less on the spot. The role of facilitator has increased my confidence. . . . As a "teller" I felt like I was the focus, and when the kids weren't understanding and I couldn't think of another way to explain it, I felt *I* had failed. With this approach I can blend into the background and yet facilitate real learning. I've discovered that the kids have better ways of explaining to each other than I ever did.

When Ginny believed that the hallmark of good teaching was knowing all the answers and being able to explain everything so that everybody always understood, she inevitably fell short. Now that she saw her role differently—to establish an environment in which her students could explore together and share their own ideas—she began to feel successful.

As Ginny's confidence in her teaching grew, she became more willing to investigate the mathematics concepts for herself. Having relinquished the belief that she was responsible for knowing the single correct way to solve any problem, she became willing to pursue alternative solutions, admit her puzzlements, and explore ideas the children came up with. In fact, as she listened to her students, her own understanding deepened.

For example, one day in January the children were busy with division problems. Ginny had given them a problem set that involved both dividing a given amount into groups of a given size (grouping) and dividing a given amount into a given number of groups (distributing). One problem required the children to find out how many floors there were in a hotel that had 240 rooms, 24 to a floor;

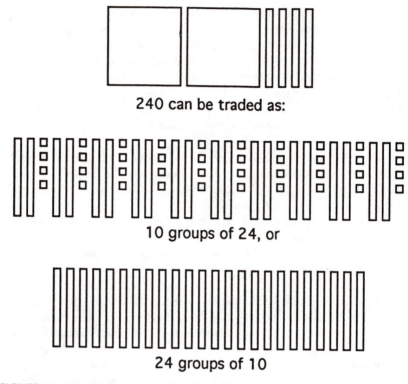

240 can be traded as:

10 groups of 24, or

24 groups of 10

FIGURE 8.1. Two representations of 240 ÷ 24

a related one concerned a 24-floor hotel, again housing 240 rooms. The children had no trouble solving the problems with manipulatives, but a dilemma arose for Ginny when she tried to help them record their solutions with pencil and paper.

For the first problem, the children who had chosen to use base-ten blocks traded in flats for rods and small cubes in order to remove successive groups of 24 (two rods and four small cubes at a time). For the second problem, they simply traded in the two flats for 20 rods and distributed the rods into 24 piles, one rod in each. (See Figure 8.1.)

"Where is the traditional algorithm?" Ginny asked when she and Cathy met after the lesson. "Look at the first problem with 24 rooms on each floor. Where's the '24 goes into 24 once'? That's the way I do division!"

Cathy got out the blocks and laid out the problem. "What does the *1* stand for in your solution?" she asked.

"Ten floors; it's one ten. But how will I ever get the kids to construct that strategy so that with pencil and paper they'll be able to use it for both grouping and distributing problems? It takes too long to keep subtracting 24s."

Cathy suggested they encourage the children to explore the relationship between the two forms of division, so they designed several additional division problems for that purpose.

When Cathy returned two weeks later, she found the children still at work on division. A few had manipulatives out, but most were simply drawing pictures to test their solutions, recording the results in numeric notation next to their pictures. Cathy asked a few students to explain.

"We're groupers, like Mrs. Brown," said Chris.

Cathy was puzzled because the problem they were explaining required distributing. Had Ginny just told them how she did division? "What do you mean?" Cathy asked.

"We discovered that it really doesn't matter how you think about the problem because they're sort of the same thing." As Ginny came over to listen, Chris laid out four rows, 31 Unifix cubes in each. (See Figure 8.2.) "See," she pointed to the length of the rectangle. "If you look at it this way, you see 31 groups of 4; but if you look at it from the short side you see 4 groups of 31." The problem required distributing 124 marbles into four jars. "We didn't want to deal out all the marbles so we just decided to see how many fours were in the 124." Chris and her partner, Dennis, had done just that, as evidenced by their drawing (Figure 8.3). They had circled three groups of four rods each, thus removing 10 fours three times, had marked these as *30,* and had then circled a single group of four units, marked as *1.*

Ginny motioned Cathy over to look at some of the other children's work and explained that the problem sets had produced a discussion about the commutativity of multiplication. Once the children had appeared to understand the connection between distribution and grouping, Ginny had told them that she found it easier to do division problems by grouping. Some had disagreed, preferring to distribute, and had called themselves "dealers." The class was fairly evenly divided between "dealers" and "groupers." The dealers' drawings for the same problem had four groups, three rods and a small cube in each. (See Figure 8.4.) Even though their drawings were different, the children seemed to understand what their answers represented.

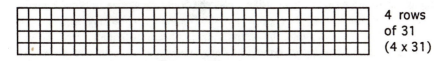

4 rows
of 31
(4 x 31)

31 stacks of 4 (31 x 4)

FIGURE 8.2. Student's demonstration of commutativity

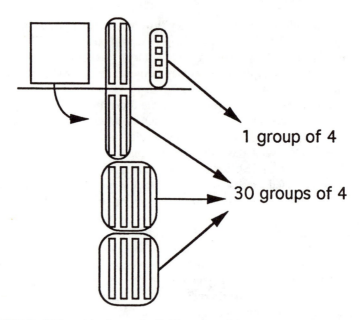

FIGURE 8.3. 124 ÷ 4 by "grouping": 31 groups of 4

In her meeting with Cathy after class, Ginny shared her excitement about
the strong connections the children were making, but was less sure about her
own progress. "I'm still struggling to understand this," she admitted with embarrassment. "I always thought of division as 'goes into' and I'm having a hard
time finding the group when I distribute. I know it works because of the commutative property—the kids helped me see that—but where's the group I'm
removing when I deal?"

Cathy responded to Ginny's embarrassment with a confession of her own.
"I learned math originally just like you, by rote." Indeed while Cathy had come
to SummerMath for Teachers with a strong background in cognitive development
and Piagetian theory, she had taken few mathematics courses in college, and
those she did take had been taught in the traditional manner. Now, working as a
follow-up consultant in the program, she often felt herself intrigued by the mathematical questions that arose for the teachers, and she could really get excited
exploring solutions with them. Like Lisa, she could recognize important mathematical questions when they arose, but she frequently had to investigate them
herself in order to understand the patterns and connections that appeared. So she
could say quite honestly to Ginny, "I enjoy investigating with you because I
continue to learn more, too. Let's get out the manipulatives and the kids' record-
ings and investigate together."

This shared investigation by teacher and consultant served to build Ginny's

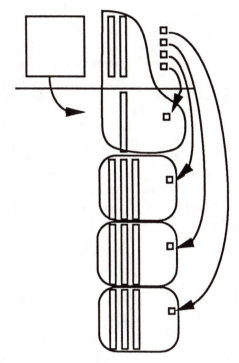

FIGURE 8.4. 124 ÷ 4 by "dealing": 4 groups of 31

confidence further. Just as Lisa's students felt empowered as they worked on mathematics that their teacher was only now coming to understand, so, too, was Ginny encouraged to learn that Cathy did not have all the answers but that they could find answers together.

As Ginny and Cathy began to work out the problem with the manipulatives and dealt one rod to each jar, it suddenly became apparent to them that one round of dealing made a group of four rods, or 10 fours. Ginny could not contain her excitement. "I understand it now for the first time! It makes sense! When I ask, 'Four goes into 124 how many times?' I'm asking either 'How many groups of four?'—that would be a grouping problem—or 'How many rounds go to four piles?'—that would be a distributing problem."

This division investigation was to remain vivid in Ginny's mind. In her journal a year later (she was then taking the mathematics course described in Chapter 4) she wrote:

My outlook on math—if not the definition itself—has changed drastically over the last year. I used to view it as an unpleasant but necessary activity.

It involved numbers, unknowns, ratios, etc., etc. I found it very difficult whether I was taking it as a course or using it in real-life situations. I was certain that unless I knew a formula I could never succeed at doing the math. It never even occurred to me that math could be understood. I slowly came to the realization that I *could* understand mathematics over the last year and a half, partly as a result of the actual SummerMath courses, and partly due to my using a constructivist method of teaching math in my classroom. My students had had this method the previous year, and they were able to teach me an incredible amount of math. I learned from them that to "know" math did not mean to be able to [mechanically] figure out mathematical answers, but rather, to understand the process, or the reasoning behind the manipulation of the numbers. I can't believe this, but I don't think I ever actually understood what long division really meant. My students—and the math approach—enabled me to understand or "know" what I was doing when I did a long-division problem. So to "know" math has nothing to do with being able to compute algorithms by using previously learned rules or formulae. It means having the ability to figure out a mathematical problem using any method or materials or diagrams, etc., that enables the person to arrive at an understanding and consequently an answer. I now view math as a challenge. I think it is fun figuring out problems.

As Ginny came to view her role as that of a facilitator of learning rather than a broadcaster of explanations, as she worked with students at their level of understanding, investigating the mathematics with them, her confidence grew. But perhaps more important to this development than her increased willingness to take risks was a different sense of what it was that one was risking. Freed from the static picture of knowledge as something one either had, and so was good, or didn't, and so was condemned, she could appreciate it as a process of which false starts and wrong turns were an inevitable and even fruitful part. The knowledge is in the process.

APPLYING THE SAME PRINCIPLES TO OTHER SUBJECTS

As the year of classroom follow-up drew to a close, Ginny began to look to the future. Although Southern Connecticut State University, where Cathy is a professor of education, is a four-hour round-trip commute from Ginny's home, she and Jill began to talk about enrolling in the master's degree program there. They wanted to extend their new understanding of the learning process to other curricular areas and to study further Piaget's work and its implications for teaching.

"This is a giant step for me," Ginny explained during her admission interview with Cathy. "Long ago I enrolled in a graduate program in geography at Syracuse University. When it came time to take the oral examination I panicked and never finished. I knew I'd flunk."

"How long ago was that?"

"Almost 30 years ago! I had finished my undergraduate work in geography and I think I was just afraid of working. And I liked going to school. The coursework was enjoyable, but I'll never forget the panic I felt over the comps. It has always deterred me from pursuing a master's degree program in education."

"Why now?"

Ginny's answer was not surprising. "I've gained so much more confidence over the last year, and I'm really excited about extending what I've learned further. Besides," she added with a smile, "there's no oral exam to take."

Together with Jill, Ginny enrolled that fall, and over the next three semesters completed her course requirements. This work guided the transformation of her instruction in language arts, science, social studies, and art. For example, one of the requirements in the language arts course was to work closely with a child throughout the semester, giving meaning to the child's writing process and reading strategies. Ginny selected Annie, one of her students, and in a descriptive case study paper analyzed the spelling rules Annie had invented in her writing:

> During the first part of this year, I am loath to admit, I taught spelling strictly with a spelling book. Although this book did provide Annie with experiences sorting words by patterns, the selection of words was not always developmentally appropriate for her. In the spring I began to use Annie's writing as a source to understand *her* invented spelling strategies. Then I designed word sorts to help her see problems in her present strategy and to enable her to construct new phonetic pattern rules. For example, Annie often used alphabetic strategies such as *lik* for *like, ran* for *rain, bik* for *bike*. Watching her sound out the phonemes as she wrote, I could see that she was using the [sound of the] name of the letter, rather than the vowel pattern, to represent the sounds she heard. I gave her a list of words, such as *rat, rate, man, main, bat, bait*, etc., and suggested she investigate the list and see if she could figure out why the words were spelled the way they were. *She* constructed the idea that the vowel sound changed depending on how many vowels there were. We then went on to investigate short vowel sounds. Working with Annie has taught me a great deal this year. I only wish I had learned it sooner.

Ginny had learned to make conjectures about the spelling rules Annie had formed, and used these as a guide to her instruction. As she had been doing in her mathematics instruction, so now she devised spelling activities that would confront her student with new information—in this case, new words—that violated her current rules. Annie's own need to resolve these conflicts would lead her to construct new rules that would account for the words Ginny had given her.

Ginny also developed a new approach to her science instruction. In a paper entitled "In the Area of Science, Can Constructivist Teaching, Based on Children's Own Interests and Investigations, Provide a Meaningful Learning Environment?" (Brown, 1989) she reported:

> I investigated with my third graders the phenomenon of light for a period of approximately six weeks. . . . The procedures involved posing what I considered to be appropriate questions, [and] encouraging the children to investigate these questions, . . . to formulate and test hypotheses of their own, to reflect on their investigations by writing and drawing in learning logs, to share ideas, to question the findings of others, and to formulate further questions which their investigations and discussions engendered. In short, children were encouraged to observe, predict, question, document and discuss their findings. (p. 1)

To begin the unit, Ginny posed a question that she thought would engage her students: *Where can you and a partner stand so that you can see each other in a mirror, but not yourselves?* After exploring this question a bit in pairs, the children lost interest and began to generate others:

> "What if we put this eraser between the two mirrors? Will we see it in both mirrors?" And "Oh, look, when the mirrors are close together, the reflection of the eraser goes on and on, but there aren't as many erasers when the mirrors are spread apart." Another pair said, "This writing is backwards in the mirror. What do I need to write so that 'Stop' shows the right way?" (p. 3)

Rather than bring them back to her original question, Ginny encouraged the children to explore their own. But though Ginny believed that student-generated explorations would be fruitful, she did not relinquish her role as guide. She periodically suggested to them questions of a more abstract nature that they probably were not yet ready to ask on their own:

> I asked them if they could tell me what light was. This was obviously something they had never thought of before, and [they] probably wondered what it had to do with the topic at hand. After a period of "think time" I received several different responses. Some said it was the light on the ceiling. Some believed it was "the opposite of dark." Others discussed sunlight, though not exhibiting a clear idea of what they were thinking. And some wouldn't venture to describe what light is. I then asked them what they thought a reflection was. Some children said a reflec-

tion is what you see in a mirror. Several said a reflection was a shadow. Although some of the children did not agree with this, they were unsure as to why at this point. (p. 3)

Ginny could see that her students were not yet ready to think about light more abstractly, but she nonetheless felt it was useful to call to their attention questions about why things happened the way they did.

Indoor mirror investigations continued for several science sessions, with children engaged mainly in "what if" types of investigations and not usually questioning "why." When I questioned them as to why they thought a specific thing occurred, the usual response was a puzzled look or a not-always-logical attempt at an explanation. (p. 3)

As the children continued their own investigations, Ginny brought in other materials—flashlights, for example—to stimulate new lines of questioning. Some students investigated symmetry and bisection of shapes using mirrors; others were intrigued by reflections and shadows; still others learned that by reflecting sunlight they could produce heat. Ginny found that the children's questions were becoming more sophisticated and their conjectures more abstract. One day when Ginny took her class outside, she recorded the following conversations:

"What if I aim my mirror at that speed limit sign; will the reflection [from the sunlight] go that far?" asked one child. She was excited that it did, and wondered aloud how far the reflected sun's rays could go. A child who overheard the question said, "I bet it could go about a hundred feet!" but another child, after several minutes of pondering, said that she thought the reflection would keep going and going until it hit something, and if there were nothing in its way, it would never stop. "But," she said, "it might get less bright."
During this outdoor session, other interesting observations were made. One boy noticed that "light reflects instantly." He said that there seemed to be no amount of time between the sun's rays hitting the mirror and then reflecting to the school building. (p.4)

By the end of the unit, Ginny had documented, through student journals and her field notes, the considerable learning that had taken place. In the last session when she again asked, "What is light?" she discovered that there had been some thinking going on about that question:

Most believed, at this point, that light moves through space at a certain speed . . . ; that it is all around us; that it reflects and is needed to make a reflection; that when it hits a mirror it bounces from the mirror to another surface. (p. 6)

And she concluded:

> I could have done a science unit on reflection and light in the traditional manner. I could have stood in front of the room and told the children all there was to know about reflection and light. I could have given them a pretest and a posttest, and perhaps found that their ability to answer questions improved, due to rote learning. This learning would have been neither lasting nor transferable. And the understandings that evolved during the science unit would not have had the opportunity to evolve. . . . The children's knowledge of light and reflection grew because they investigated these topics in a manner which was meaningful and relevant to them. (p. 7)

Ginny published her paper in *The Constructivist,* the newsletter for the Association for Constructivist Teaching (ACT) (see "Resources for Teachers" in the Appendix). In it, she included this insight:

> It was more difficult for me to pose probing questions during these science activities than it was during math—probably because I understood the science less and had never actually worked with reflections, mirrors and light myself. Since I discovered [this approach to] teaching, I have also discovered how much I don't know, never even realizing it was there to be known, but learning it along with the children. (p. 4)

Two years earlier, not knowing had been a shameful fact to be hidden. Fear of being found out had kept Ginny quiet and withdrawn. Now she could freely discuss what she had learned from her students in a newsletter read by several hundred educators.

SPEAKING UP

Each risk taken helped strengthen Ginny's confidence further. When she agreed, together with Jill, to lead a workshop for teachers, administrators, and teacher educators at the annual conference of the ACT, she was petrified. But the workshop went well and the feedback was enthusiastic. Later, she co-led four workshops for teachers in her district.

Rather than these workshops, though, another incident stands as most vividly representative of Ginny's transformation. At that same ACT conference, with 200 people assembled for a panel discussion on constructivism, one of the panelists, an ACT board member, was addressing the phenomenon of cognitive disequilibrium. He had invited discussion of how the area of a rectangle outlined by string would compare to the area of the parallelogram that would result if the string were pulled slightly at two diagonally opposite corners. A number of participants in the audience thought the area was conserved. Ginny raised her hand. "No," she declared in a firm voice so as to be heard clearly. "The area is decreasing. If you continue pulling at the corners, the parallelogram will collapse to an area of zero."

The panelist did not acknowledge Ginny's point, and someone else offered another opinion: "The length stays the same, but the height is decreasing. Since area of a parallelogram is height times length. . . ." Now the panelist was nodding in agreement.

"But you don't even need to know a formula to plug in!" Ginny interrupted, realizing that her solution had not registered. "Don't you see? Just stretch the string to its limit. It quickly becomes apparent that the area is approaching zero."

In front of 200 people, most of whom were strangers, Ginny was holding her own. With her newfound confidence she was explaining her mathematical reasoning with assuredness and clarity, even though she knew that many in the audience had stronger mathematical backgrounds.

Two years later, when Cathy was preparing to write this chapter, she and Ginny discussed the incident at ACT.

"What were you feeling that day?" Cathy asked.

"I knew my answer was a good one. But I felt like he was looking for *his* answer, and that made me angry."

"So you trusted in yourself so much that you were willing to argue with him in front of 200 people?"

"Yes! Once I reached my own conclusion, I knew it was logical, and then I felt confident." Ginny paused a moment and then went on. "That's why this whole thing—the program, teaching this way—has been so important to me. It changed me, and I'm not a person who likes change. I get set in my ways and enjoy the security it provides. But I liked this change I saw in myself, in my classroom. It was painful, but wonderful. In the beginning, at the first institute, I hardly said anything. Later I felt stronger mathematically, and I think I began to see mathematics differently."

"What do you mean?"

"Well, I don't feel like there is only one right answer now, that I can be judged by if I don't know it. And this is tied into my whole change of view regarding mathematics teaching. I'm not looking for one right answer. Instead the kids brainstorm and offer multiple strategies which we all discuss together."

"Do you remember how nervous you were about my coming into your classroom once a week?"

"Yes!" Ginny smiled. "I'm not so nervous about people coming in anymore. In fact, I invited my principal in the other day. I felt comfortable because I don't think there is one right way to teach anymore, either. I don't feel like someone is 'up there' looking down at me critically."

Ginny was quiet a moment before she continued:

Teaching this way has changed my life. I hadn't been doing much in the way of professional development, partly because of the fear of failure. Now I've built a confidence that I never thought I could have, making me want to try new things in other areas as well. In this approach there is such freedom to really learn.

When Ginny discarded the related myths of an all-knowing Math God and of mathematical knowledge as a perfected "it" that the brilliant somehow have, but that she never would, she found herself free to enjoy her own learning and that of her students. The risks no longer seemed so risky, nor failure so final— she had come to see knowledge as an open-ended process.

9
Learning to Lead: Geri Smith

Although I have studied about how people learn, I was surprised to rediscover that the learning process is the same for an individual regardless of age. . . . The teacher role of assessing where the learner is presently, of posing questions or situations which reach to the edge of the learner's understanding . . . is very much the same for eighth-grade students, for third-grade students, and for peers (teachers).

Marie Appleby
South Hadley (MA) Middle School

Previous chapters have followed the varied journeys of four teachers who envisioned for themselves a new mathematics instruction and then set to work realizing that vision in their classrooms. Their success in transforming their practice—in some cases they were alone in their schools in attempting this—should not obscure the larger truth that classrooms are not autonomous units. Teachers must constantly negotiate the many, and sometimes conflicting, forces that impinge upon them from without. As they work to achieve a reformed mathematics instruction, their efforts may be either enhanced or diminished, depending on whether—or the extent to which—the demands of administrators, parents, school committees, colleagues, textbook publishers, politicians, and the media are consistent with their own goals.

In this chapter, the focus widens beyond the classroom to include the school and its context.

Just as individuals develop beliefs, habits, and assumptions about teaching and learning that form the basis of their pedagogical decision making, so, too, school communities develop, through time, well-established mores affecting school routines, expectations, and policies. In general, traditional school cultures both create and reflect the assumptions that in the classroom children learn only from their teachers, that learning is a matter of accumulating facts and practicing appropriate procedures, and that good teaching is measured by the number of right answers students "get." These assumptions govern decisions—usually administered down through a power hierarchy from superintendent to principal to teacher—on such issues as how to assess student learning, which textbooks to use, and how to evaluate classroom instruction. The most common elementary-school scenario features a single teacher, brought together with approximately 20 to 30 students of common age and ability in a self-contained classroom, following directives from principal, textbook, or other authorities. Other adults rarely enter the classroom. When they do, it is usually to evaluate the teacher. Collaboration with peers is uncommon and what happens inside the classroom is rarely shared with colleagues.

The pressures on teachers who work to enact new instructional paradigms within the traditional school culture can be intense. Teachers must justify their practice to skeptical parents, supervisors, and colleagues; they must balance their own beliefs about the mathematics their students should be learning with the need to prepare them for standardized testing; and they must do all this largely on their own, with little or no intellectual or moral support from other members of the school community.

The mathematics education reforms being called for will be assimilated more easily if groups within that community, rather than just lone individuals, begin to challenge existing school culture. Because many assumptions will have to be "unlearned" and new ones constructed before profound or widespread change can occur, that change will come slowly (Joyce, 1990; Sarason, 1971, 1990). But only through such a process can the culture take on the reforms as its own.

This analysis figured among the reasons why SummerMath for Teachers began to develop options for those among its participants who wished to become educational leaders. Under one grant some led workshops in their districts, and under a second, some assumed classroom follow-up responsibilities for their colleagues. In both cases, the intention was to provide contexts in which teachers could reflect together on classroom process while introducing into the school culture more sustained and mutually supportive collegial relationships.

When Geri Smith entered the program, she was able to transform her quite traditional mathematics instruction with relative ease. But the story this chapter tells is how she learned that the principles of learning and teaching that had come to guide her work with her fifth and sixth graders were equally valid for her work with her colleagues: they had to be given the opportunity to construct new principles of pedagogy much as her students were now being given the opportunity to construct upper-elementary mathematics. In following Geri's story, one begins to see the elements of an emerging culture committed to school- and even district-wide change.

TRANSFORMING CLASSROOM INSTRUCTION

Geri had always been a leader. She was "head teacher" in the elementary school in which she had taught for 13 years. This meant that, in addition to teaching her fifth- and sixth-grade classes, she was in charge when the principal was out of the building. For years she had served as the K–6 curriculum coordinator, participating in that capacity in district meetings on mathematics and science. And even before entering SummerMath for Teachers, Geri had often conducted workshops for her colleagues, bringing back to them what she had learned in the various in-service programs she attended. In fact, just before her first SummerMath for Teachers institute, she had been funded to teach a summer "computer familiarization" course to her elementary colleagues.

On her application to SummerMath for Teachers, Geri wrote:

> For several years I've been looking for something to rejuvenate my mathematics program. It's important to me to keep my curriculum current and to continue to discover and use different ways to present the material. . . . My experience for the past twenty years has been in grades two through six. . . . In addition to this experience, I would bring enthusiasm and an eagerness to revitalize my program. I have been in many positions of leadership throughout my years of teaching and I would look forward to working with other teachers if given the opportunity.

The two-week SummerMath for Teachers institute was "exhilarating" but "painful" for Geri. She was comfortable with problem solving, and so she enjoyed the mathematics activities and felt excited about her solution strategies. But a pedagogy informed by constructivist perspectives contrasted sharply with one based on principles of reinforcement, external motivation, teacher explanation, and drill—all elements that had been central to her teaching. As Geri reflected on the excitement she had felt about her own mathematics learning in the institute, and on the instructional principles that had made it possible, she felt confused and insecure. In her journal she wrote:

> We brainstormed [in my group] to find solutions, and, [through] working and building on each other's ideas, what seemed at first to be a most puzzling task became exciting and fun. I had a good "safe" feeling as I approached the solutions and loved having others to bounce ideas off of. I also like the idea of finding more than one method of solving each problem. Not only did I feel that my ideas could be "right" but I was encouraged to keep thinking after one solution was discovered. . . .
>
> But I'm having a problem accepting the fact that it is better for a teacher not to tell a child his effort is "right," but rather to comment on his approach only. . . . I feel that a positive comment to a child will encourage him to want to continue and possibly have more success. . . .
>
> I've had a lot of confusion in my mind this week as to the teaching method used. . . . [In a listening activity today] I had a difficult time suppressing the need to lead the learner. I will have to free myself of the responsibility I feel for getting the learner to my answer.

These issues continued to preoccupy Geri for several days, and she often discussed them with fellow participants. Journal entries suggest that at the core of her puzzlement was a dilemma over the goals of instruction: was the point of her teaching to lead her students to "the right answer" or was it to help them construct their own mathematical understanding? By the end of the institute, she had resolved that dilemma for herself:

> After these weeks of being a learner, I am much more aware of a learner's needs. The anxieties, successes, failures—[the] emotional roller coaster I've been on—have made me much more sensitive to the child. This increased sensitivity has changed my belief about *how* children learn. They don't learn best when they are regurgitating what I've taught them, but rather when they teach themselves. . . . Reinforcement can be a hindrance in that it may end the thinking process and not encourage the child to explore other possibilities. . . .
>
> The most obvious thing that I didn't seem aware of before is that we need, as teachers, to leave the method of solution open to the learner and not stifle their learning by presenting one solution or inferring that there is only one way. I also never really explored why or how children reached a solution as long as the answer was "right." . . . What I had thought of as part of math (problem solving) I realize now is in fact the essence of teaching math.

The mathematics explorations that she had enjoyed so much had also given her a new sense of mathematics itself—one that aligned it more closely to her experience of natural science.

> I hadn't made the connection before today, but many of the science classes I teach begin with an open-ended problem to solve. By experimenting with materials, the group reaches a solution which is later shared with the class. I question why I do this in science and not in mathematics. Is it that I'm aware that science is always changing—[that] new discoveries are continually being made? Have I been teaching mathematics as a stagnant discipline? I suppose I was taught math as a method toward solution to be memorized—facts and algorithms unchanging and coming to us from some unknown deity never to be tampered with!

At the end of the institute, Geri shared with staff a set of personal commitments for the coming school year:

> My commitment to myself is to try to implement the following techniques in my teaching: use manipulatives to help learners understand, visualize,

and explain their thinking; use word problems appropriate for the learners' ages and interests; use small groups to work out problems cooperatively and large groups to share findings and consider all possibilities; use probing questions to stimulate original thinking and to help children explain their procedures [so they will understand] them better themselves; emphasize concepts over mechanics, concentrating on understanding rather than on memorizing algorithms. Although this is an ambitious list, I'm very excited about implementing each technique. I am aware, though, that this implementation will be gradual.

The following school year, Cathy worked with Geri to help her achieve these goals. They began in September by concentrating on class management issues, but because Geri had such wonderful rapport with her students, these concerns dissipated quickly. Once the lessons were going smoothly and groups were working well together, most conference time was spent planning activities to develop conceptual construction.

Though Cathy worked closely with Geri, it was Geri's classroom, her curriculum, and ultimately her decision which changes she would try to make. Geri usually introduced the lesson to the children; then she and Cathy would move from group to group, listening and questioning. Geri took the lead in facilitating whole-group discussions (which Geri, Cathy, and the class termed "math congresses"), but Cathy would jump in when it seemed appropriate.

Around the beginning of November, feeling that she just couldn't figure out how to begin the unit on double-digit multiplication, Geri asked Cathy to lead the initial lesson. Aware that the children had already encountered the traditional algorithm, Cathy decided to develop their understanding of the operation by focusing on the idea that the algorithm is based on the distributive property. She began by telling them a story (a variation of the Xmania story found in Chapter 3):

> Once upon a time, long, long ago, people only knew how to multiply big numbers one way. If they wanted to know 15 × 13, they wrote *13* down 15 times and then added it all up. This took a long time to do. Luckily for the people, there was a wise old inventor who discovered a new way! Unluckily, she died before she could explain her idea. But she left some materials in her hut with a note that read, "I have invented a short-cut system to multiply big numbers. These materials gave me the idea."

Holding up wooden base-five blocks, Cathy invited the class to use them to figure out a shortcut to the answers for 7 × 5, 12 × 5, 17 × 5, and 26 × 5.

Some children began by laying out seven rods, but usually a partner would

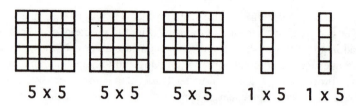

FIGURE 9.1. Students' representation of 17 x 5

say, "That's the long way!" Others quickly set to work with the flats and large cube. Geri and Cathy asked the children to record their solutions with pictures and numbers so they could later share them with the class. The diagram in Figure 9.1 is representative of the work of most groups and served as a focus for the whole-class discussion.

Brydie spoke for her group. "We think that you multiply out all the five-by-fives first. Then do the one-by-fives that are left over. That's what we have here with the blocks." She laid out 17×5 using three flats and two rods. "That's $75 + 10 = 85$."

"That rule is just like ours," Tony said excitedly.

"Help me understand why it works," Cathy requested. "I'm not sure I'm convinced this rule will always work."

"Yes, it does," Tony responded quickly. "See, $5 + 5 + 5 + 2 = 17$. You've done them all. Try another—26×5 is $(25 \times 5) + (1 \times 5)$ because $25 + 1 = 26$. And you can even do bigger numbers in your head, like 53×5." He thought for a moment and then announced proudly, "It's $250 + 15$."

"What if you wanted to multiply 12×13?" Cathy pushed them beyond the base-five manipulatives. "What different ways could you multiply to get the answer?"

Several solutions were offered: $(6 \times 13) + (6 \times 13)$, $(5 \times 13) + (5 \times 13) + (2 \times 13)$, and $(10 \times 13) + (2 \times 13)$. The children were developing their own algorithms.

When Cathy and Geri met after class, Geri expressed her excitement about the lesson and began to plan how she would continue teaching multiplication using other base materials. But she, too, needed to understand more clearly how the multiplication algorithm builds on distributivity, wanting to see it visually represented. "I have to see for myself where the traditional algorithm is with the base-ten blocks. I don't see it." Geri spread out the base-ten blocks and then represented 12×13 as in Figure 9.2.

"Where is the 2×3?" Geri asked with exasperation. "I still don't see it."

"Why do you want to find 2×3?"

"Because I want to multiply 2×13, then 10×13 like in the traditional algorithm."

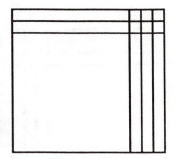

FIGURE 9.2. 12 x 13 with base-ten blocks

Cathy gently separated 2 × 13 from the larger array. "Then let's look at the 2 × 13 first."

"There it is!" Geri was excited now as she decomposed the matrix into its traditional component parts (Figure 9.3). When she had finished, she emphasized the importance of providing the children with a similar experience. "They're just like me," she explained. "They learned that algorithm by rote last year. I'll be interested to see if they can find the parts after we work for a while with other base materials."

Geri believed that her students needed to construct their own algorithms, but she also wanted them to understand the meaning behind those they had previously encountered. So she planned the next several lessons with these goals in mind, and as the weeks went by, and they used both traditional algorithms and their own inventions, the children's thinking about multiplication became very flexible.

By February, Geri had made terrific strides toward realizing the goals she had set for herself the previous summer. She was excited about her mathematics classes, and so were the children. Now other teachers in the school were becoming curious, often stopping in during Geri's math period and even inviting her into their own classrooms, asking her to teach demonstration lessons.

Geri's main concerns now were with lesson planning. Although she routinely began a new unit with a series of word problems, she was still unsure about how to design and sequence problems to promote concept development. A lesson on fractions was instrumental in helping her gain insight into this issue.

"*I* ate more," Alex was declaring as Cathy arrived for her weekly classroom visit. Geri had asked the children to figure out who had eaten more clams: Alex, who had had ⅜ of a bucket, or Charlie, who had eaten ¼. The children had made their own sets of fraction bars by taking paper strips of equal lengths and folding and cutting them into various fractional parts, and Alex had placed a ¼ strip next to three ⅛ strips to show that ⅜ was more. "See. The ¼ is smaller than the ⅜."

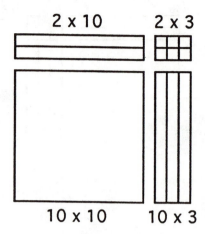

FIGURE 9.3. 12 x 13 in traditional component parts

"We did it a different way," Sara responded. "Adam and I lined ours up like this and we think they are the same." Sara and Adam had positioned their three ⅛ pieces atop one another horizontally instead of vertically (see Figure 9.4) and were judging quantity by the height of the two columns.

"But that ⅜ is fatter," argued Alex.

"So we could just cut some off."

"But would it still be ⅜?" Geri asked.

"Yes," responded Adam. "'Cause it's still 3 out of 8 parts."

Geri and Cathy looked at each other, amazed at the apparent misconceptions. Cathy thought she saw what the students might be struggling with, so she decided to explore their concepts of the whole. She took one set of eight eighths and lined the pieces up vertically and then placed another set, lined up horizontally, next to it. (See Figure 9.5.)

"Are these wholes equal?" Cathy asked the class to vote. To the surprise of both teachers, everyone voted "no"!

Then, slowly, a couple of the students wearing puzzled frowns began to change their votes. "Yes they are. One is just fatter, so it doesn't go up as high. It's still the same amount."

"If you cut some of the width off and put it on the top, you'd see. It would be the same."

"But Adam and Sara said before that they would just throw the extra away," Geri reminded them.

"I think you have to put the bit you cut off on the top, otherwise it's not the same whole."

The debate continued for the rest of the period and into lunch, with children arguing over what the whole was and what a fraction was.

When Geri and Cathy met after class, Geri was elated: "I think that's the best lesson I've ever had! Did you see the puzzled looks change with the dawning of understanding? They were so involved! I know there's lots more discussion and manipulation of materials needed, but now I know where to go next. We'll continue discussing this tomorrow and cut up graph paper to check out their theories. Then once their concepts of fractions are solid we'll explore equivalents, then operations. If you start with problems and listen and probe for their understandings, then the unit just evolves. They develop the unit with their own questions!"

The following summer, during the advanced institute, Geri might have had this lesson in mind as she described her current conception of teaching:

> To encourage and facilitate learning, the teacher must provide meaningful activities. Although it may be difficult to find the solid core of knowledge in a student, it is from there that the teacher should proceed. By presenting appropriately meaningful problems that are thought-provoking, the teacher begins the process of providing for learning to take place. If the

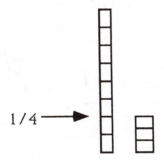

FIGURE 9.4. Students' demonstration that 1/4 = 3/8

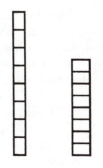

FIGURE 9.5. Two arrangements of 8/8

problems are carefully planned, they will take the learner from one discovery to another. They will create a state of disequilibrium so that the learner will be compelled to search for a solution. It is the teacher's responsibility to be sensitive to the frustration level of the student. I like to compare this state of disequilibrium to a boat sailing along on calm waters. It is not until the water is churned up a little that the boatman has to work. In this turbulence he learns what to do to keep an even keel. As teachers, we should provide problems that will encourage our students to action and be wary of the hurricane waves that will overturn the boat.

This description of the teaching process was based on a very different set of beliefs than the one Geri had espoused just one year earlier. She was now convinced of the need for students to be able to argue, ask questions, and explore mathematics at their own levels of understanding. The teacher was responsible for facilitating these activities, providing a well-thought-out sequence of problems in whose solution the children were invested.

A NEW CONCEPTION OF LEADERSHIP

For Geri and the other teachers who volunteered to disseminate to their colleagues, through in-service workshops, what they had learned in the program, the challenge of becoming an in-service leader again "churned up the waters," providing more "turbulence that would require work to keep an even keel." Although the teachers had examined the learning process and had begun to reorganize their own classroom instruction to take account of what they had discovered, they had not considered whether the principles that they were now applying in teaching their students ought also to govern their work with adults. But they soon found it necessary to rethink their model of the workshop leader as someone who presents new ideas, explains them, and then provides a fun activity for participants to try out in their classes the following week. They now had to ask themselves what they wanted to accomplish in these workshops, which ideas they wished to communicate, and how best to do so.

Geri thought seriously about these questions. After one advanced institute session devoted to them, she wrote in her journal, "What can we tell about [this approach to instruction] and what must others discover or experience? I'm hard pressed to find ideas I can tell." Later in her journal she touched on the issue of ownership, having played a game with herself as an experiment:

> In a recent discussion I decided to try to listen but not participate verbally. I found it frustrating to hold back, but even more interesting is the problem I had trying to stay involved with what was being discussed. This

experiment made me realize that [when I was not] committing myself to the problem by sharing ideas, my mind easily wandered. I felt no ownership for what was being discussed. I must remember this.

And she did. In subsequent planning sessions, Geri was adamant that workshop participation be interactive and voluntary. She argued:

It's when people experience mathematics that it becomes more real to them. I know they are going to want us to tell them how to teach math. But I think they need to experience learning mathematics so that they can reflect on what learning mathematics is all about. Then, as teachers, they can be reflective decision makers, not just cookbook followers. You have to experience learning mathematics in order to make decisions about how to teach it. Learning should determine the teaching, not teaching the learning. And, if the workshops are voluntary, the teachers are there because they want to be.

With these ideas in mind, her workshops, co-led by Cathy and one of Geri's colleagues, were similar in format to the lessons described in Chapters 2, 3, and 4. Teachers worked in groups, using manipulatives, to explore adult-level mathematics embedded in real-life problems. Following the mathematics activities, workshop participants were encouraged to identify and reflect on the pedagogical assumptions that had organized the lessons. Sessions usually ended with discussion of how these principles might be applied in elementary-school teaching.

While participants responded positively to the workshops, as evidenced by their evaluation questionnaires, the impact of these few sessions on classroom instruction proved negligible. The kind of transformation envisioned was simply too profound to be sustained on the strength of three or four two-hour workshops. (For example, Lisa Yaffee [Chapter 7], who had attended a similar series of workshops and had been inspired to introduce problem solving into her instruction, found that she had been given no guidance in how to use this technique to teach the mathematics she was responsible for.) Thus, although the workshops did provide a forum for discussing alternative classroom process, their impact did not, except in isolated cases, extend beyond the two-hour session.

In themselves, the workshops did not—could not—create classroom, let alone schoolwide, change, but a number of participants did become interested enough in what they had glimpsed to subsequently enroll in a SummerMath for Teachers institute or spring course. (Linda Sarage [Chapter 4] decided to enter the program after attending a workshop led by Geri.)

As the workshop project approached its end, SummerMath for Teachers staff analyzed its results and designed a successor project to overcome the first project's limitations and build on its achievement. Under a second NSF grant, a project was established in which teachers who had demonstrated particular aptitude would be selected to provide the kind of classroom follow-up for their colleagues that program staff had hitherto been providing. The "resource teacher internship" was designed with several goals in mind: to extend beyond the reach of program staff what had proven the single most effective catalyst of pedagogical transformation; to offer at least to some teachers the extraordinary opportunities for learning that visiting a variety of classrooms on a regular basis made possible; and to help build collaborative relationships among teachers so as to counteract the isolation that is so common in the schools.

To enable Geri to become her district's resource teacher intern, her superintendent relieved her of half her teaching load, freeing her for follow-up work. As part of the grant-funded activities, Geri also attended a weekly seminar, led by Deborah Schifter, for all teachers providing follow-up for colleagues.

The first meeting of Geri's seminar group began with each of the interns being invited to talk about the "autobiographies of change" they had prepared for the session. They had been asked to address such questions as: how did you feel in September when you first began to make changes? what were your goals at the time? what realizations did you make during the year? what were the events that stimulated those realizations? what was your relationship with your classroom consultant? what did he or she do that you found to be helpful? This activity was meant to prompt the interns to think about their new role as classroom consultant; and, so that they might come to see the variety of forms that role might take, they were asked to share their conclusions with one another.

Deborah wanted the interns to appreciate that their charge was to learn about the goals, concerns, and interests of the teachers they were working with, and, based on that information, to help those teachers reflect on classroom process. In contrast to more conventional mentoring programs, the resource teacher intern was not expected to provide solutions to problems or tell the teachers what to do. Instead, he or she was to support the development of a practice suited to each teacher's vision while suggesting possibilities for consideration.

When the school year began, the interns did not fully understand that the process, as well as the result, of change would have to be individual. Once program staff realized this, they deliberately chose seminar topics and formats that would help interns develop greater sensitivity to the individuality of the teachers they visited. For example, interns would role-play teachers they worked with while staff would interview them. Afterward, the group would discuss such questions as: what does this teacher believe about how children learn mathematics? what is she trying to teach? and what are her immediate goals? Such exercises prodded the group into thinking in terms of teachers' agendas and, at the same time, forced them to clearly articulate different ways of thinking about learning and classroom practice.

Reflecting on these issues, Geri wrote in her journal:

At first I thought I had figured out the right way to teach because everything had gone so well in my own classroom. I thought that my role was to help teachers do what I did. I thought they were supposed to run their classes the way I ran mine because it was right. . . . I knew what had happened with me when I changed. I thought it would be the same for the teachers I'm working with. What I realize now is that it is so different for everyone. The process of change is so individual. It depends on the teachers' background, their experiences, their style, on how they work with students.

At times, interns brought to a seminar session issues that had become immediate concerns in their schools—when is it appropriate to expect students to learn their multiplication tables? what roles can textbooks play? how does one deal with skeptical parents? At other times they came with mathematical questions or asked for help with ideas for lessons. Most often, however, discussion revolved around what was happening with the teachers the interns visited.

As Geri worked to understand how teachers learn and how they transform their own pedagogy, her conception of her own role began to change. In one seminar session, she talked about two teachers in her school, both of whom taught very conventionally. They assumed that they were the rightful focus of attention in their classrooms, and the way the children's desks were set up—in rows facing front—reflected that assumption. "It's so tempting for me to tell them to arrange the desks in a circle or a U-shape to facilitate more interaction among the children," Geri explained. "If I tell them to do that, I'm sure they would, because supposedly I know something that they don't. But I want them to do it because *they* feel the need for it. I think my job becomes trying to create that need."

A week later Geri reported that she had succeeded in doing just that:

When I met with one of those teachers after class, I asked if she thought the children were involved in the discussion. When she said, "not as much as I would like," we brainstormed ways to facilitate more interaction. First she said that she might call on more children to respond. So I rephrased the question to emphasize encouraging the children to *want* to be involved. So now she began to talk about how, when children were behind others, they couldn't hear. That's when I suggested we try changing the arrangement. *She* felt the need to try it.

Geri went on to explain that she had had a similar conversation with the other teacher. "One teacher changes the arrangement only for math—she rearranges after math class—but the other teacher has kept the desks that way now for all subjects and has commented that she sees the children more involved and learning from each other."

Later in the year, she reported another example illustrating how important it is not to simply give directives, but rather to encourage teachers to reflect upon their practice and make decisions for themselves. In this case, Geri was working with a second-grade teacher:

> She had asked her class to find how many children there were in the classroom if a parent brought in 36 cookies but discovered that 12 more cookies were needed so that every child could have 2. Some children were claiming that they added; others said they subtracted. They were all right! Their explanations matched what they had done in action with cubes to find the answer. Some had added 36 + 12 first, then divided by 2; others had used repeated subtraction and removed groups of 2 from 36, then groups of 2 from 12. The teacher commented to me that she was really rethinking her past practice of telling kids "key words" to look for in word problems. . . . I could have told her not to teach key words like "all together," "in all," and "difference" in the beginning of the year, but it wouldn't have meant as much to her. Now she sees for herself why teaching those words with an operation can be more confusing for children.

As the weeks wore on, Geri's ideas regarding her work with colleagues crystallized as she connected that process to her experience of analyzing learning. "What I'm finding," she explained,

> is that it's almost like déjà vu. It's like what I discovered with the children learning mathematics. If I tell the teachers what to do or what to look for, it's not as dynamic or rewarding for them. They don't have any ownership of it. [When I've been a student] the best teachers I've ever had [were] not the teachers who have lectured. They were teachers who worked with me. That made it fun. They invited me to participate with them in solving math problems, in writing, in investigating science. I hope I can be like that and that I can help other teachers be like that.

In her mathematics instruction, Geri had learned to listen for and recognize the validity of her students' own perspectives, helping them to discover their own approaches to problem solving. Now Geri learned to do this for her colleagues:

> Eventually I realized that their classrooms didn't have to look like mine, but it could still be right—for them. It's almost like that perfume that was on the market which claimed to have a unique fragrance depending on the person who wears it. [Even when teachers were working from a similar set of goals and principles] I could see how different it looks on each

teacher. This helped me to encourage the teachers to be who they are and to incorporate the ideas of how people learn that they've acquired through their SummerMath experience into their own style.

Geri also learned that her role could shift depending on the strengths, interests, and concerns of the teachers she was working with.

Some teachers . . . were eager to try some of the new strategies they had learned about and while working on one strategy would smoothly move on to another. Each obstacle they encountered created a new situation in which to learn more. I also observed the difficulty other teachers had in changing what they were doing in the classroom. They were insecure about doing something different and clung to what was familiar.

For some teachers, Geri could be a sounding board, helping them reflect upon the impact of their new strategies on student learning. For others, it was important for Geri to be encouraging, to make them feel safe and help them find challenges that they would see to be within their reach. "I'm also finding," she reported, "that some teachers who I thought would never change, can change. I just have to be patient."

During Geri's first summer institute, three years earlier, when she had begun to analyze her own mathematics learning, she had recognized the importance of owning what she had learned, of constructing it for herself. After successfully implementing changes in her teaching practice, she volunteered to introduce what she had learned to her colleagues. She soon realized that just as her students could not share *her* experience of mathematics but would have to construct their own, so her colleagues could not share her new understanding of teaching but would have to invent one for themselves. In Geri's words, "they need to own the change."

For Geri, the key to fundamental and enduring change was an understanding of the nature of the learning process. The more she learned about the process of construction, the more committed she became to instructional reform. Each new challenge, each new role, brought her experiences that broadened and deepened this understanding.

Geri's developing understanding of the learning process provided the framework for decisions she made as classroom consultant, but her work as consultant also yielded, reciprocally, new insights about learning. A year after she had become a resource teacher intern, she described this connection:

It was learning about learning that was the most exciting, scary, wonderful part for me. And it's still going on. Every teacher I work with has a different strength and I learn so much from them. When you're in a mentor role

where you're not directly responsible for children, you can take a moment
and step back and really observe what is going on—how the teacher's approach affects the children, how the style affects the children. Some teachers have such a wonderful style and the children are so wonderful in return. I love to see that and then also how a teacher's attitude affects the
relationship with children. I can see so much that I couldn't see before
when I was too close. You also learn a lot from the children. You really
see the progression of concepts being developed—all in the same year.
I've taught second and third grade before, but it's not the same as seeing it
all happen at once. It's so exciting. And I see where the weaknesses are. I
see teachers teaching place value [in] second grade and still [in] the sixth.
I'm seeing a good overall view of what happens in mathematics education
in the elementary school. I'm still learning.

CREATING A CULTURE OF REFORM

Seymour Sarason, in his book *The Predictable Failure of Educational Reform* (1990), argues that for change to occur, "schools must be reconceived as
places that exist coequally for the development of students *and* educational per-
sonnel" (p. 145). Because of the work Geri and her colleagues were doing together, their elementary school was becoming just such a place. It was no less
invested in the development of students—one could argue that it had become
even more so. But it had clearly also become a place for the development of
teachers. In the old kind of place, teachers were supposed to come with all the
answers. But now, Geri and her colleagues could unashamedly say, "I'm still
learning."

They were learning mathematics (many were attending courses offered by
the program) and they now sought one another out in school to work on problems together. And they were learning about their students' learning, together
devising new activities to facilitate construction. Geri reported:

Doors are opening. It used to be that everyone taught behind closed doors
and nobody ever had much of an idea about what would go on behind that
closed door. But it's not like that anymore. People keep their doors open
and they'll sometimes pull another teacher in to see what's going on. And,
you know, people never used to talk to each other about what they were
doing. It used to be that in the teachers' room people would hardly ever
talk about their classes, unless it was to complain about one particular
child. Now they talk about what they're trying. Or when two teachers
pass in the hall, it's likely that one will stop the other and say, "Listen,

I'm having trouble with such and such a concept. What do you think?" That's really different.

Wanting to capitalize on the energy being generated by the teachers, the superintendent of schools established a K–12 mathematics curriculum committee, appointing a number of SummerMath for Teachers participants to sit on it, and asking Geri to chair. His goal was a new curriculum to legislate and drive reform.

But after an initial meeting, the committee realized that teacher development was the key to the changes the district was after. "We decided to focus our attention on teacher change, to work towards educating teachers, rather than writing a new curriculum," Geri explained. "Because then the teachers in turn will want to change." Though he had had other ideas, the superintendent concurred. The committee then formulated district needs and goals in relation to in-service and decided to set up both parent and teacher workshops.

This event inaugurated a change in the district's policy-making structures. When the teachers on the committee determined that the task they had been charged with—to redesign the mathematics curriculum—was not their highest priority, they decided on another. They were in the best position to judge their colleagues' needs and claimed the authority to act.

Another example of teacher-driven policy change concerned tracking. To people who think of knowledge as information transmitted by someone who possesses it, ability grouping makes sense. Their assumptions are that, first, schools exist so that the well-informed teacher may convey to large numbers of children what they should know, and, second, instruction must necessarily be pitched to one particular level at a time. Otherwise, clearly, students above that level will be bored and students below it won't understand. Thus goes the common argument for tracking.

After participating in SummerMath for Teachers, a number of teachers from different districts began to dispute that conventional wisdom. Deciding that they prefered "mixed-ability" grouping, some teachers in a federally funded program for low-achieving students chose to work with all the children, collaborating with the regular classroom teacher, rather than work alone with the children specifically assigned to them. Other teachers had come to believe that their high-achieving students learned more if they stayed with their classmates to explore grade-level mathematical concepts further, rather than separating themselves in order to race through the textbook to get to the next grade level (Riddle, 1991).

In Geri's school, students had traditionally been tracked, beginning in grade one, into "high" and "low" classes based on test scores. However, after participating in the program, most teachers had grown dissatisfied with this arrangement. Finding themselves in agreement, they proposed to the superintendent of schools that it be eliminated. Again he accepted their suggestion. After one year, all the teachers in the school, whatever the group they had previously taught, reported that the new system provided a much richer mathematics environment.

Teachers in Geri's school continued to meet both in and out of committee, exchanging views about teaching and learning, discussing issues that arose in instruction, and working together on mathematics. Where once teachers had worked in isolation looking for right answers and clear explanations, they now valued collaborative inquiry and reflection—for both their students and themselves. Traditional structures of authority were beginning to concede to them some ownership of the process of transforming instruction. A new culture of reform was emerging in the school as teachers worked together to reconstruct mathematics education.

10
Conclusion

I say look to your colleagues. We're the ones reinventing mathematics education.

Lisa Yaffee
Marks Meadow Elementary School
Amherst, MA

The teachers whose stories we tell in this book are probably not typical of all of the many thousands who are, or soon will be, confronting the challenge of transforming their mathematics instruction. They are not even typical of the 500 or more who have participated in SummerMath for Teachers. Yet in choosing them as subjects of this book we were guided by our belief that, during this period of ferment, providing compelling portraits of individual possibility is at least as important as formulating cogent statements of principle.

However, since the success of the current wave of reform requires that *large numbers* of traditional classroom teachers confront critically the habits and assumptions that now guide their instruction, it must be convincingly shown that teachers can, in significant numbers, successfully transform their practice; and, if they can, what, realistically, such a transformation entails. In this concluding chapter, we draw upon the history of SummerMath for Teachers in order to address these questions directly.

CAN TRADITIONAL TEACHERS TRANSFORM THEIR INSTRUCTION?

The experience of SummerMath for Teachers shows that coherent, long-term teacher development programs, programs that have thought through and internalized the meaning of the new paradigm of mathematics instruction, can effect the kinds of far-reaching changes widely called for. However, we must emphasize that this broad claim does not exclude the possibility, indeed the certainty, that many teachers will be unwilling or unable to make such changes—even when these programs are designed to acknowledge and address the numerous difficulties that will arise.

Some, like Elizabeth, who was very active and involved throughout the summer institute in which she participated, will ultimately decline to compromise smoothly running classrooms. Elizabeth's written work evidenced serious reflection on how learning takes place and what kind of classroom might best facilitate it. But finding herself back with her first graders in September, faced with the dramatic contrast between familiar classroom structures and those she had envisioned during the institute, she panicked and withdrew from the program within six weeks of the start of classroom follow-up.

Others, like Jane, who has generally been regarded as the best mathematics teacher in her primary school—her efforts confirmed by her students' test scores—will remain convinced that their instructional approach is fundamentally sound. During her 20 years as a teacher, Jane has kept her top students challenged by taking them rapidly through the grade-level curriculum and beyond and has sent those who could not keep up to other teachers. Though she was intrigued by the challenging mathematics problems she encountered in a SummerMath for Teachers course, 20 years of affirmation had validated her practice.

A third, larger group of teachers, of whom Kara is representative, will return to their classrooms with some new instructional techniques, but will use them in the spirit of a traditional practice. Kara, who had never been particularly strong in mathematics, realized during a course she took with us that she herself learned more working cooperatively and using manipulatives. However, the conclusions she drew from these experiences never touched her basic assumptions about learning and about mathematics.

Still others will find themselves unable to relinquish the role of "resident genius." Joe was one of these: after attending a summer institute and while receiving follow-up support, he became enthusiastic about the way diagramming word problems had enabled him to understand rules he had previously taken for granted. Now he was anxious to share his insights with his students, but instead of allowing them to explore the fractions problems for themselves, he lectured about them, solving them at the blackboard as he talked.

Teachers like Marta, who are intrigued by the ideas offered by the program, will remain unconvinced that their students can learn except through constant drilling. Others, like Felicity, complain that they have to work so hard to make up interesting problems that the pressure of nightly preparation simply becomes too much. And still others—RuthAnn, for example—tire of continually defending themselves against parental skepticism.

Elizabeth and Jane represent the small number of teachers who come to us for whom the fact that a constructivist perspective on classroom process differs so substantially from their own is itself sufficient ground for rejecting what we have to offer.

Far more numerous are those, like Kara and Joe, who incorporate new strategies into their practice, but only in the most obvious and superficial ways. While such teachers often report renewed enthusiasm for teaching and are excited about their own learning, they never confront the deeper issues we raise.

Whether they will ever allow themselves to entertain serious doubts about the habits and assumptions they cling to so tenaciously is an open question.

Finally, there are many teachers like Marta, Felicity, and RuthAnn who *are* interested in refashioning their mathematics instruction, but who are unable to overcome what *they* see as insuperable barriers to change. Such teachers need far more support than they—often rightly—feel they receive, and have given up, at least for a time. But were they encouraged by what we have called a culture of reform, the chances are very good that these teachers would try again.

Though we have only briefly touched on our experience with those teachers who reject the new paradigm out of hand, or who interpret what we offer as a set of trendy strategies, never examining how learning takes place, it is important to acknowledge that many such teachers exist. The kind of teacher development SummerMath for Teachers has sought to promote is possible, but it is not easily achieved. If we do not believe that this admission defeats our claim that teachers can, in significant numbers, successfully transform their practice, what evidence do we have beyond the stories told here?

An Assessment Instrument

In order to have a basis for making judgments about SummerMath for Teachers participants in the aggregate, we interviewed, each June from 1986 to 1991, all teachers who had just completed a year of classroom follow-up support. The interviews were analyzed to assess teachers along two axes of innovation: "implementation of particular strategies" and "instruction based on a constructivist epistemology" (see also Schifter & Simon 1992; Simon & Schifter, 1991).

"Implementation of particular strategies" refers to whether, as a result of their participation in the program, teachers introduced any new techniques into their classroom repertoire. Because we felt that it was important that teachers select strategies that made sense to them in their own teaching contexts, we did not specify which these should be. From our work in their classrooms, we knew that most elementary teachers had introduced manipulatives, group work, and/or nonroutine problems.

In order to summarize comprehensive data in a way that takes account of the developmental nature of the implementation process, we borrowed the Levels of Use (LoU) instrument created by G. E. Hall and his colleagues (1975) at the University of Texas at Austin. On the LoU rating scale, teachers move from not knowing about an innovation (level 0, "nonuse"), to learning about and preparing to use it (levels I, "orientation," and II, "preparation"), to using it with different degrees of sophistication (levels III through IVB).

At level III, "mechanical use," teachers have started employing the innovation, but are preoccupied with problems of classroom management—for ex-

ample, the most efficient way to distribute manipulatives or how to keep student
noise levels down. At level IVA, "routine use," teachers are no longer struggling
with management, and implementation of the innovation runs smoothly. At level
IVB, "refinement," teachers begin to adapt the innovation to meet the specific
needs of their students. An important shift takes place between levels IVA and
IVB involving fuller internalization of the innovation by teachers and a corre-
sponding change of focus from their own teaching behaviors to student needs.

In the LoU interview, teachers are asked to describe how they employ the
innovation, what they see as its strengths and weaknesses, and what plans they
have for its future use. Developmental level is ascribed by the interviewer based
on the teacher's description of his classroom practice. The rating assigned is the
one achieved for whichever strategy was used most frequently and comfortably.

While it is important to know how well teachers are using new instructional
strategies, the implementation of this or that technique does not, in itself, indicate the development of a new instructional practice. This point is vividly made
in the following excerpt from an institute journal:

> Prior to my experiences here, I felt children could be successful learning
> math as a result of a hands-on approach. I felt that since I was fostering
> active learning, the children really could understand what they were
> doing.
>
> I understand now that although my classroom instruction provided
> for cooperative learning and "appropriate" learning activities, there was
> little emphasis on thought. This realization really pulled the rug out from
> under me.

What this teacher highlights is the difference between implementing specific teaching strategies and operating out of a particular structure of beliefs
about learning and knowing—an epistemological perspective. That is, any strategy can be employed to help students remember, get right answers, or have fun,
without regard for the kinds of understandings being constructed. For this reason, and from the inception of the SummerMath for Teachers Program in 1983,
staff understood that it had to look beyond teaching strategies to those beliefs
about knowing and learning that are enacted in classroom practice. In 1986, this
understanding was codified in an instrument called the Assessment of Constructivism in Mathematics Instruction (ACMI).

Like the LoU, the ACMI consists of an interview and a procedure for ascrib-
ing a rating. Yet, while the development of a teaching practice based on constructivist principles usually involves changes in some fundamental beliefs about
learning and teaching, teachers are not asked to spell out those beliefs directly.
For what we want to assess is not how well they can articulate philosophical
perspectives or how closely their vocabulary matches our own. Instead, we want

to determine the nature of those beliefs about learning and teaching that guide teachers in the context of the instructional decisions they make.

To get at the ways in which teachers are thinking about the learning process, we elicit from them information about what they want their students to learn and what considerations enter into decisions about the nature and organization of instruction. Discussion focuses on specific examples—for example, how was instruction on a particular topic thought about, planned for, implemented? Responses to questions such as these reveal the theories of learning that teachers enact in their classroom practice.

We chose to employ a scale similar to the LoU to assess "instruction based on a constructivist epistemology," but with significant alterations. The original LoU scale envisages a pattern of implementation that progresses from mechanical use of a prescribed behavior (level III) to ownership of the ideas behind the behavior (level IVB). However, changes in one's conceptions of learning and their subsequent application do not (at least in our experience) tend to follow this pattern. Rather, what we typically observe is that participating teachers leave the courses or institutes with rudiments of a constructivist epistemology—some general and quite abstract views on learning—and then set to work putting those ideas into action.

At ACMI level 0, there is no evidence that constructivism plays any role in instructional decisions. Teachers' descriptions of their own practice imply a conception of student learning as a process of information reception reinforced by drill and practice. Teachers are responsible for explaining clearly what students should know and, like Kara or Joe, when they adopt new strategies, these are subordinated to traditional instructional paradigms.

At ACMI level III, teachers evidence rudimentary understanding of constructivism, but it finds no effective expression in practice. For example, a teacher has come to feel that it is important for students to construct their own mathematical understandings. So, rather than telling them what they should be learning, she asks them to solve some problems in the hope that they will come up with the right conclusions on their own. At this point she has not yet developed a scheme comprehensive enough to suggest where this process of construction might best start, nor how to structure students' work for optimum learning. And she is most likely struggling to reconcile her new ideas with old curricula, textbooks, and so on. In short, she is preoccupied with changing her instruction and does not yet know how to make student learning the basis of her instructional decisions.

In her interview at the end of her first year of participation in the program, Sherry described how she had come to realize that her students had to put ideas together for themselves, that her explanations rarely had the intended effect. She wanted, she said, to provide opportunities for her students to solve problems and make connections, but she was still struggling to transform this aspiration into a coherent practice. Although she had, by then, a number of classroom strategies in place (using nonroutine word problems for students to solve in groups with manipulatives), she was assigned to ACMI level III.

At ACMI level IVA, teachers have become comfortable with more active student involvement and commensurately less teacher telling. Their teaching behaviors are now consistent with their views of a transformed practice, but their attention to students' cognitive constructions remains subordinated to their concern with their own behavior. Had Jill been interviewed that first October after attending the summer institute, she would have been assigned to level IVA. For at that time, while she was excited about and comfortable with the new strategies she had developed to promote student learning, she was still more attentive to her "approach" than to individual students' developing constructions.

At ACMI level IVB, teachers are no longer preoccupied with their own actions, concentrating instead on student learning. They are able to monitor student understanding and can fluently revise their lessons as necessary, adding intermediate steps, extending applications in one direction or another, or confronting misconceptions. In short, teachers at ACMI level IVB base their decision making on what their students understand, and as a result are much less concerned with what they themselves should or should not be doing. The focus is

Figure 10.1   LoU and ACMI Levels

Implementation of Particular Strategies: LoU

Level 0: does not know about or has rejected use of the strategy.
Level I: is learning about the strategy.
Level II: is preparing to use the strategy.
Level III: uses the strategy; struggles with problems of classroom management with
   respect to the strategy.
Level IVA: implements the strategy with ease.
Level IVB: fine-tunes strategy to meet the specific needs of students.

Instruction Based on a Constructivist Epistemology: ACMI

Level 0: does not hold a constructivist epistemology.
Level I and II: not applicable.*
Level III: has a rudimentary understanding of constructivism, but has difficulty basing
   instruction on this understanding.
Level IVA: holds a constructivist epistemology and is comfortable with instruction,
   but focuses on teaching behaviors.
Level IVB: focuses on students' learning from a constructivist perspective.

*Since the ACMI was used to assess the extent to which actual classroom *practice* reflected a constructivist epistemology, levels I and II, which in the LoU concern *learning about* an innovation, do not apply.

Table 10.1.   Summary of LoU and ACMI Data After One Year of Program
Participation

	LoU (strategies)		ACMI (epistemology)	
Level	#[a]	%[b]	#[a]	%[b]
0	4		46	
III	42	(97%)	29	(66%)
IVA	43	(66%)	22	(45%)
IVB	47	(35%)	39	(25%)

[a]number of teachers at that level
[b]percentage of teachers at that level or higher
Based on 136 interviews of elementary teachers between the spring of 1986 and the spring of 1991.

on student learning in particular—not just on teaching behaviors that generate student-centered lessons. As an example of instruction guided by and responsive to student learning, recall Lisa's description of her fractions unit, taken from an ACMI interview and recounted in Chapter 7. There she describes how she designed problems that would facilitate further construction based on what her students' explanations revealed about their understandings.

Figure 10.1 summarizes the LoU and ACMI levels.

LoU and ACMI Results

Table 10.1 shows data based on interviews conducted with all elementary teachers who completed the classroom follow-up component with either SummerMath for Teachers staff or resource teacher interns between the fall of 1986 and the spring of 1991.

LoU results indicate that 97% of the teachers who completed classroom follow-up implemented strategies (such as group problem solving or use of manipulatives) at level III or higher. Level IVB, which indicates not only stable use but willingness to adapt strategies to learners' needs, was achieved by 35% of the teachers.

According to the ACMI results, 66% showed evidence of at least a rudimentary constructivist view of learning as the basis for their teaching (level III or above), while 25% were facilitating student construction by focusing directly on learning (level IVB).

The LoU and ACMI data confirm the importance of distinguishing between those whose learning was restricted to the acquisition of new teaching strategies and those whose views of mathematics learning and teaching shifted fundamentally. Not surprisingly, innovation in teaching strategy was more easily and rapidly achieved than were changed views about learning as enacted in instruction.

In analyzing the results of these interviews, it must be borne in mind that they were conducted after only one year of participation in the program. The stories of Jill, Sherry, Lisa, Ginny, and Geri—and others we might have chosen to tell—illustrate that development may continue well beyond the follow-up year.

While we do not have extensive numerical data, we did interview one set of 15 teachers who attended the advanced institute and then conducted in-service workshops for colleagues during the year that followed their own year of classroom support. Table 10.2 displays the results of interviews conducted in consecutive years with this group of teachers. The data indicate both greater facility with particular strategies after two years (93% at level IVB, up from 60%) and a practice more substantially grounded in a constructivist epistemology (87% at level IVB, up from 40%).

Admittedly, this group was self-selected—they had chosen to attend the advanced institute and volunteered to conduct in-service workshops—and no control group was interviewed. It is therefore not possible to gauge the effects of a second year of program participation by comparing the group that continued to participate with an equivalent group of teachers sharing the additional year of experience but lacking the additional year of participation. However, it is worth noting that over 50% of the teachers who have participated in classroom follow-up have taken at least one additional course with the program, though some wait a year or more before they feel ready to do so.

Assessing the ACMI

In 1986, in conformity with guidelines set out under its first National Science Foundation (NSF) grant, SummerMath for Teachers designed an instrument to assess the effectiveness of its work. In fact, staff had always been guided by an understanding of the distinction between implementing particular strategies and making instructional decisions based on a constructivist epistemology. But devising and then employing the ACMI proved an important exercise in self-clarification: in sharpening that distinction, and especially in identifying the stages through which teachers typically pass as they develop a practice based on constructivist principles, the ACMI defined the program's identity. More recently, however, as this manuscript neared completion in early 1992, and with the concept of the big ideas of the mathematics curriculum moving to the center of our account of the new instruction, this definition has come to seem too narrow.

Specifically, as Chapter 5 passed through successive drafts, the impact that Jill's work on "the missing addend" has had on her practice came to seem a more and more important part of her story. What it was telling *us* was that teaching for mathematical understanding meant teaching for the construction of the big ideas. We had always intended that Jill's story would serve to plot the typical

Table 10.2.   Comparison Between One and Two Years of Participation

	Year 1				Year 2			
	LoU (strategies)		ACMI (epistemology)		LoU (strategies)		ACMI (epistemology)	
Level	#[a]	%[b]	#[a]	%[b]	#[a]	%[b]	#[a]	%[b]
0	0		2		0		1	
III	2	(100%)	3	(87%)	0		1	(93%)
IVA	4	(87%)	4	(67%)	1	(100%)	0	(87%)
IVB	9	(73%)	6	(40%)	14	(93%)	13	(87%)

[a]number of teachers at that level
[b]percentage of teachers at that level or higher
Based on 15 teachers, each interviewed after one and two years of participation in SummerMath for
  Teachers

trajectory of teacher change—in effect, putting the ACMI levels in motion. The stages of Jill's development, "First Faltering Steps," "Initial Innovations Become Routine," and "Shifting the Focus to the Learner," were to correspond to ACMI levels III, IVA, and IVB, respectively. But now we argued that only after she had learned to base her instruction on close attention to her students' understandings had Jill begun to see what they should be constructing. "Finding the Big Ideas for Second Grade" seemed an additional, necessary, and crowning stage of her journey to a mature and coherent practice. But in bursting the confines of the ACMI framework, this story also revealed ACMI's one-sided character: missing from the chapter was an analysis—as opposed to an anecdotal account—of Jill's development as a mathematical thinker and how that had prepared her to identify and teach for the big ideas.

   In our work with teachers, we had always acted on the belief that for them to achieve a successful practice grounded in constructivist principles required a qualitatively different and significantly richer understanding of mathematics than most possessed. Especially important was the development of a different conception of the nature of mathematics, both for gaining deeper insight into the learning process and for permitting a more sophisticated approach to mathematics content. If, for analytical purposes, we distinguish between an "epistemological" and a "mathematical" strand of the teacher development process, we can say that we have had no theory either of the mathematical strand or of its relation to the epistemological. For us, the concept of teaching for the construction of the big ideas now poses the challenge of theorizing the unity of the two strands as a single complex process.

   How do we characterize and assess the process by which teachers develop a level of mathematical understanding adequate for facilitating student construction of the big ideas? And how can we help teachers learn to think about mathematics in terms of the big ideas of the school curriculum? What are those ideas? Can they be identified by an analysis of mathematical structure without reference to children's cognitive development? Or is there an essential role here for empirical research? These are some of the questions we hope to pursue as we continue our work.

Acknowledging these limitations of the ACMI framework is not inconsistent with our continuing reliance on it. At this point in our own development as teacher educators and researchers, the framework still seems to us an indispensable tool for thinking about and assessing a crucial aspect of our work.

This being said, a final caution: not all teachers follow the course plotted by the ACMI levels. For example, when Lisa entered the program, her instruction in subject areas other than mathematics, where her understanding was limited, reflected a well-developed constructivist perspective. However, soon after the start of the school year, her expanded view of the nature of mathematics and her increased confidence in her mathematical abilities enabled her to create a program consistent with her constructivist commitments. She did not need to wrestle with contradictions arising between old and new paradigms: in short, she skipped level III. And, because she knew to attend to the progress of her students' constructions, she skipped level IVA as well. Already that first year, Lisa's work involved exploring the big ideas of the upper elementary curriculum.

Hugh, another teacher who has worked with us, though he is not otherwise mentioned in this book, entered the program with a very rich understanding of high school mathematics. In his many years of teaching he had identified those concepts with which his college-bound students struggled year after year as the crucial ideas they needed to grasp. But until he realized that his students' understandings could become the basis of his instruction, he had found no alternative to patiently repeating his elegant and lucid explanations.

Though Lisa and Hugh did not conform to the ACMI pattern, in fact they are exceptions that prove the rule. In our experience, teachers who have as sophisticated a grasp of constructivism as Lisa, or are as mathematically developed as Hugh, are very rare. Most have had little or no experience with constructivism and possess a limited understanding of mathematics. Though Jill's development was remarkable for its rapidity, it was typical in those respects.

WHAT DOES SUCH CHANGE INVOLVE?

As our discussion of the ACMI's limitations points up, our work leaves many questions unanswered. However, much has been learned about what will be required to successfully transform mathematics instruction along the lines of the new paradigm.

Developing an instructional practice based on constructivist views cannot be simply a matter of adopting a different textbook, introducing manipulatives, or putting students into small groups, for all of these things can happen within the frame of a pedagogy that centers about training students to get the right answers to problems of given types. In addition to the shift in epistemological perspective, constructing a practice consistent with the new paradigm entails a new conception of the nature of mathematics and a drastically revised picture of what should be happening in the classroom.

Most teachers who come to us have never before thought about the questions we pose for them; the idea that teaching should be grounded in an understanding of *how students learn* and of *what they should learn* seems obvious only in retrospect. For these teachers, as for their peers, learning to teach has been largely a matter of learning how to maintain discipline and present materials devised by "experts." Most professional development involves exposure to new lesson ideas and new techniques for broadcasting them. As one teacher admits, mid-institute, in this journal extract:

> It is really difficult to discuss how my beliefs about the nature of math and the learning of math have changed because, in all honesty, it really isn't something that I have focused on. . . . Pressure has been on how well [my students] compute—how many areas . . . I've covered. . . . The math director says [to] focus on problem solving, use manipulatives—and we think about and discuss what *methods* work, but not really on: they do or don't work because *this* is what math is and *this* is how we learn.

At first, teachers are disconcerted that we ask such questions at all. This is not what they came for, we are frequently told; they entered the program to learn how to *teach* mathematics, not philosophize about it. However, as the power of these questions becomes apparent, many begin to recognize their implications for classroom practice.

While visions of possibility can take shape in an introductory course or summer institute, the bulk of what teachers must learn will necessarily come only in their own classrooms with their own students. The new mathematical understandings they must develop and the teaching situations they will have to negotiate are too varied, complex, and context-dependent to be anticipated in one or even several courses. As teachers begin to change their practice to reflect new insights into the nature of mathematics and mathematics learning, they will be confronted with issues and ideas that could not have been predicted. Thus, just as Lisa characterized her insight that learning mathematics is "learning how to ask questions and think about the questions you ask," so, too, the same can be said of learning about learning and learning about teaching.

We have at several points emphasized that the activities designed for SummerMath for Teachers courses and institutes aim not so much at covering some necessary minimum of material but rather at inaugurating a complex process of active reflection—on what is involved in serious exploration of mathematical concepts, on how to think about student learning, and on the advantages and limits of different kinds of classroom structures. This process, if sustained, becomes the core of a new practice.

Even veteran and conventionally successful teachers find that each mathematics topic taught must be understood in new ways. For example, in Chapter 8 we described how Ginny, no longer satisfied with mechanically transmitting the division algorithm, felt it necessary to explore different interpretations of that operation; and in Chapter 9, we saw how the Xmania exercise gave Geri an experience of mathematical understanding that served her as a model when she came to examine the relationship between a physical representation of the product of two two-digit numbers and the conventional multiplication algorithm.

As teachers like Ginny and Geri wrestle with essentially the same mathematics content their students are working on, they are also learning what to listen for in what their students are saying. What mathematics are these students constructing? If they are confused, where are the wrong turns? What is the logic in this child's misconception? What question can I ask to help him break through his limited understanding?

The teachers learn that some confusion is idiosyncratic and some of minor importance. But many confusions and no longer serviceable conceptualizations are widely shared and will not be cleared up by a simple intervention. For example, year after year, Jill sees how her second graders struggle with the idea that a missing addend can be found by subtracting. Similarly, Lisa's fifth and sixth graders had to work long and hard to understand how 1/4 of a cake can, at the same time, be 1/2 of an uneaten portion. Typically, such big ideas of the elementary curriculum as that addition and subtraction are inverse operations and that a given object can be represented by different fractions, depending on what the reference unit is, emerge only when teachers give their students time to work through problems that raise these issues.

Teachers who make these changes may find themselves preoccupied with issues of classroom management—how to impress on their students new expectations about working in groups, handling manipulatives, and so on. But once new routines are settled, other problems arise. Whereas in the traditional classroom almost all communication—and authority—centers about the teacher, now she must establish a flow of ideas among her students and between herself and her students in both directions. The success of this kind of classroom structure depends on teachers and students learning to listen to one another and to act cooperatively.

Not surprisingly, once the transition has been weathered, many teachers report that their students become more socially mature, better able to share and take risks, increasingly reliant on one another rather than on the teacher. But the more complex social environment that results from the dispersal of authority also requires greater alertness to those gender, race, class, and ethnic dynamics that contribute to exclusion from mathematics-related fields. When teachers loosen controls over student interaction, patterns of marginalization may actually become more apparent than when students sit quietly while their teachers talk. Who speaks up in discussion, whose ideas are pursued in small groups, who handles the manipulatives, and who sits at the keyboard are matters to which teachers must learn to attend. Lisa decided that she had to take action— by forming a girls' math club—in order to alter patterns of communication in her classroom and give the girls the confidence to speak up. Where such solutions are inappropriate, teachers must find other creative ways to address analogous issues. Ensuring the full participation of students whose first language is not English in a setting heavily dependent on spoken language is a particular concern of teachers with whom we work.

The teacher development effort described in this book can be characterized as a three-fold commitment: helping teachers to construct an epistemological perspective informed by the principles of constructivism, offering them opportunities for new and deeper understandings of the mathematics they teach, and supporting them as they develop a classroom practice guided by these epistemological and mathematical conceptions. For most teachers, this process of transformation is highly affect-laden—human beings are not just cognitive devices—and Sherry's pain, Lisa's anger, and Ginny's anxiety are more the norm than the exception. Our experience suggests that those who work in teacher development must be prepared to see teachers through—not help them evade— discomfort, confusion, even despair. After all, teachers are being asked to face up to the disparities that emerge between their new, evolving conceptions of mathematics instruction and their own past habits and assumptions, and then to abide the uncertainties that attend the translation of these vaguely defined conceptions into a confident and effective practice. But feelings are not incidental inconveniences on the way to a better teaching practice; curiosity, pride, and affection and concern for their students are among the reasons teachers keep at it. It helps immeasurably if they learn to identify what they are feeling and consequently understand the flow and quality of that feeling as predictable and necessary aspects of the process of change.

Given the complex and far-reaching character of the kind of transformation described here, the commitment to a coherent teacher development process must evidently be long-term. Significant instructional change takes time, reckoned in years rather than months, even with considerable support and opportunity for reflection. At SummerMath for Teachers, the most effective form of support has proven to be weekly classroom follow-up by a staff member or resource teacher intern. But even those teachers who seem readiest to take advantage of what is offered find that, at best, the year of follow-up defines an agenda for further development. Additional avenues for pursuing that agenda, in advanced institutes and semester-long mathematics courses, for example, are vital; teachers need and are eager for the structure, stimulation, and collegiality they provide.

And they are thirsty to explore mathematics, welcoming courses that provide opportunities for investigating mathematical concepts with their peers. There they can develop a more expansive picture of the mathematical landscape, beyond the materials appropriate for their own grade levels. Yet traditional mathematics courses of the sort offered at most colleges and universities do not satisfy this thirst. Linda's and Lisa's experiences are very revealing on this score: first, an investigation of elementary-school topics, such as operations with fractions and decimals, can be appropriate and challenging for adults; second, exploration of such topics provides context for the development of "mathematical thinking"—making conjectures, proving them to oneself and others, articulating generalizations, and so forth; third, such explorations also provide occasions for reflection on the processes of learning and teaching mathematics; and fourth, the fact that the main topics of discussion are drawn from the elementary curriculum does not preclude discussion of what are generally considered more "high-powered" topics, such as limits or non-Euclidean geometry. Linda showed us that at the same time one is learning "basic" facts (for example, that the angles of a regular hexagon do not change as the sides of the hexagon grow), one can also inquire into the role of mathematical convention in making meaning, or appreciate the drama of introducing a set of principles incompatible with Euclidean geometry. Mathematics explorations based on the elementary curriculum but designed for adults can be every bit as demanding and conceptually sophisticated as calculus. In earlier chapters we described how our courses, institutes, and workshops give teachers the opportunity to discover their own powers of mathematical thought.

Another important lesson we have learned is that instructional development in individual classrooms will necessarily be limited unless school cultures change along with them. Many schoolwide policies—assessment of student learning, evaluation of teaching, or curricular decisions, for example—directly affect what can happen in the classroom. Teachers and administrators working together, and reaching out to their communities, can establish new policies that support the kind of instruction proposed by the reform movement.

As this chapter is being written, SummerMath for Teachers is in the second year of a third NSF-funded project, the Mathematics Process Writing Project. The many teachers who have worked with us to successfully transform their mathematics instruction have acquired an invaluable fund of experience, and this project was designed to encourage them to make that experience available to their colleagues. Each year, 14 to 18 teachers write about the problems and dilemmas that arise as they work to change their classroom practice. Their papers capture the process of instructional decision making from the perspective of the teacher, and bring to life convincing depictions of the powerful mathematical thinking that can take place in a class of typical second, fifth, or tenth graders.

In one of these papers, Lisa Yaffee (1991) has pungently summarized many of the points made in this book:

> It seems almost coincidental that the National Council of Teachers of Mathematics has recently come out with standards that affirm the ideas we have been struggling to implement in our quiet corner of west central Massachusetts. We know that teachers all over the country . . . have been working to create a meaningful context in which their students can learn and use powerfully the web of concepts that is mathematics. It's time for us to share what we've learned in the interest of promoting universal math equity and literacy. Everybody, not just the lucky few with "math minds," should be able to feel good about themselves as math problem solvers. No one should have to suffer the foreclosure of options engendered by not knowing how to think mathematically. Math is not arcane. Anyone can learn it. Anybody can teach it too, but not until they confront the gaps in their own understanding.
>
> For those of us who learned by the book, the process of confrontation might require throwing the textbook away and turning to our colleagues for help. We are all smarter as group members than we are as individuals, and right now the group of teachers interested in increasing the accessibility of mathematics to everyone is growing. . . .
>
> I say look to your colleagues. We're the ones reinventing math education.

But is Lisa's conclusion—that it is she and her colleagues who together are "reinventing math education"—more than mere hyperbole? After all, policy makers and administrators, textbook and software writers, education researchers and teacher educators are the ones used to being listened to when "expertise" is called for. Indeed, when learning is understood as information acquisition and teaching as information transmission or relay, situating the creative work of education reform outside the classroom has a certain plausibility. Since "experts" on curriculum can decide the content, sequence, and manner of presentation of mathematical topics, teachers need not grasp the underlying logic of curricular organization—its big ideas—or concern themselves with models of learning and teaching. It would seem a corollary of this view that the relevant theoretical assumptions can be "hard-wired" into instructional materials that even the worst-paid and least-prepared teacher can employ successfully. In mediating between the experts and their students, teachers need only open the box and add water.

By contrast, the new paradigm for mathematics instruction can be enacted only when teachers themselves grasp the big ideas, internalize the models, and then put them into play. For example, as teachers come to understand the big ideas that underlie the mathematics they teach, they also realize that they cannot simply "transmit" these ideas to their students. Instead, they must provide opportunities for their students to explore and to wrestle with concepts as they arise. Such a practice cannot rely on predetermined scripts but depends on one's capacity to respond spontaneously to students' questions and discoveries. This is one of the insights that Lisa's use of the concept of "invention" captures: for her the term implies a process open to the unforeseen, an outcome not predetermined. While teachers must design activities to address particular issues of content, they must also cultivate the virtue of opportunism, taking advantage of openings even at the expense of departing from short-term agendas.

As the locus of pedagogical authority shifts from experts and administrators to classroom teachers, those teachers must become more dependent on one another. With greater autonomy comes greater responsibility for structuring the learning process. Teachers will need opportunities for collective reflection in order to make considered and effective instructional decisions. When Lisa exhorts teachers to turn to their colleagues for help, it is because they best understand the contexts that call for those decisions.

Yet we would extend Lisa's exhortation: *all* of us working for mathematics education reform should be turning to the teachers, for it is they who will invent the practice that realizes the vision. Only through their work can we come to know, concretely, and at its best, what mathematics education can be.

Appendix

RESOURCES FOR TEACHERS

Teachers, teacher educators, and administrators interested in further exploring applications of constructivism to the classroom can obtain information through the Association for Constructivist Teaching. Write to:

Catherine Twomey Fosnot
ACT
The Center for Constructivist Teaching
Southern Connecticut State University
New Haven, CT 06515

For information about the SummerMath for Teachers Program, write to:

SummerMath for Teachers
Mount Holyoke College
South Hadley, MA 01075

A CHRONOLOGY OF SUMMERMATH FOR TEACHERS

1981   Mount Holyoke College hired Jere Confrey to design and direct a summer mathematics program for high school–aged women. Jere brought to the SummerMath Program a grounding in contemporary research in mathematics education and gathered a staff inspired by this opportunity to create an alternative mathematics pedagogy.

1982   Jere collaborated with Jack Lochhead's Cognitive Processes Research Group (CPRG), based at the University of Massachusetts, Amherst, recruiting a number of her summer staff from there: Jack, Jof Clement, Peter Rosnick, and Ron Narode. Linda Cavanaugh from nearby Greenfield Community College and Richard Upchurch, who was then at a community college in South Carolina, also joined the summer program staff at that time, as did Deborah Schifter.

The first six-week SummerMath Program for high school–aged women was conducted in the summer of 1982. (A similar program has taken place every summer since then.)

1983   One year later, Jere and her colleague, Joan Ferrini-Mundy, who was then on the faculty of Mount Holyoke College, received a one-year grant to conduct summer institutes introducing teachers to some of the principles that formed the basis of the SummerMath Program. That summer Joan directed two institutes for high school teachers called SummerMath for Teachers.

1984   Several new people joined the SummerMath community: Ann Hickey, who taught in the high school program, and Fadia Harik, who directed the institutes for teachers. (Joan had by now gone on to the University of New Hampshire.) Marty Simon and Gini Stimpson were hired to work in both programs.

When Jere and Fadia left Mount Holyoke at the end of the summer of 1984, Ann became the director. That fall she contacted administrators and teachers from three local school districts to initiate a discussion of how SummerMath for Teachers could serve their schools.

1985   Marty joined the discussion that spring when he was hired to direct SummerMath for Teachers.

The result of these discussions was the Educational Leaders in Mathematics (ELM) Project, a two-level program. At the initial level, teachers who attended a two-week introductory summer institute received weekly clinical supervision during the following academic year. About half of these teachers went on to an advanced level comprising a second institute and an apprenticeship program in which they learned to conduct workshops for their colleagues.

New staff were hired for the SummerMath for Teachers institutes: Ellen Davidson, Paula Hooper, and Alison Birch.

During the first year of the ELM Project (funded by Title II and the Geraldine R. Dodge Foundation), Marty, Ellen, and Deborah made up the team of follow-up consultants for the 14 participants.

1986   When the ELM Project (now funded by the National Science Foundation [NSF]) expanded to 30 participants the following year, Cathy Fosnot, a professor at Southern Connecticut State University (SCSU), joined the staff as a part-time consultant. Also at this time, Deborah became assistant director of SummerMath for Teachers.

Ann left her position, and Jim and Char Morrow were hired to co-direct the program for high school women.

1987   Virginia Bastable, a high school teacher and past participant in SummerMath for Teachers, joined the staff. Cathy took a year's leave of absence from SCSU and worked in the program full-time.

1988   Jim Hammerman joined the SummerMath for Teachers staff. Cathy returned to her position at SCSU but continued part-time on the project.

When the ELM Project concluded in 1988, Marty left for a faculty position at Pennsylvania State University, and Deborah became director. At the same time, SummerMath for Teachers was awarded a new NSF grant, the Mathematics Leadership Network (MLN). MLN introduced two new components into the program: semester-long mathematics courses that meet during the academic year and a resource teacher internship program.

The primary objective of the semester-long MLN courses was to give teachers an opportunity to explore in depth the mathematical concepts of the school curriculum.

The MLN resource teacher internship program was developed when it became clear that it was the follow-up component that was crucial to the success of ELM and that training outstanding teachers to do classroom follow-up, rather than conduct occasional workshops, would more rapidly and effectively produce district-wide change. Through MLN, selected past participants were supervised and supported as they learned to provide that follow-up. The internship program—by replicating the ELM model, but staffed now by teachers—moved districts toward continuing development of their mathematics programs with minimal support from SummerMath for Teachers staff.

1990 Jill Lester (Chapter 5), a second-grade teacher and past program participant, joined the summer institute staff.

The program received a third NSF grant for the Mathematics Process Writing Project (MPWP). Under MPWP, each year 14 to 18 past program participants are invited to write reflective narratives about aspects of mathematics teaching, and especially about their instructional decision making. In support of the reflective writing process, the teachers attend a semester course and three day-long workshops fashioned after the "process writing" model. A principal aim of the writing project is to provide the kinds of materials needed by teachers and teacher educators who are attempting to align their practice with the NCTM Standards.

1991 MLN was completed and the MPWP writing courses began.

SummerMath for Teachers was awarded a grant form the Dwight D. Eisenhower Mathematics and Science Education Program for a project entitled Improving K–8 Mathematics Instruction Through a Focus on Geometry.

1992 The institutes will continue into the foreseeable future. In addition to federal funds, SummerMath for Teachers has received support from a number of private sources: the Geraldine R. Dodge Foundation, the Pew Charitable Trust, and the Rockefeller Brothers Fund. Current staff include Deborah Schifter (director), Virginia Bastable (assistant director of the program and mathematics teacher at Amherst Pelham Regional High School), Paula Hooper (a graduate student with the Media Project at the Massachusetts Institute of Technology), Jill Lester (who continues to teach second grade at Westhampton Elementary School), Marsha Davis (a member of Mount Holyoke's Department of Mathematics, Statistics and Computing), and Ricky Carter (of Bolt, Beranek, and Newman).

THROUGHOUT THE HISTORY of the SummerMath for Teachers Program, staff have shared a commitment to mathematics instruction and teacher education, but each has come with a different background and an interest in particular aspects of the project. We have advanced degrees in mathematics, psychology, computer science, and education. Though we have all been teachers, some of us have taught special needs students, others have had high-achieving classes, and still others mixed-ability groups. We've taught in elementary, junior high, and senior high schools (as well as colleges and universities), and in public schools, private schools, and alternative programs. Thus, whenever we sit down to plan workshops or institutes, discuss the teachers with whom we work, or prepare for program assessment, our individual perspectives are inevitably brought to bear. Yet far from being a source of difficulty, this diversity has contributed immeasurably to our personal and professional development, even as it has created a rich collective identity for the program.

Over the last several years, we have become part of an informal network of teacher educators whose mathematics in-service programs are premised on similar goals and commitments. In addition to those who have been on our staff and are now at other institutions, these educators include Lynn Hart and Karen Schultz at Georgia State; Carolyn Maher, Bob Davis, and Alice Alston at Rutgers; Paul Cobb and Terry Wood at Purdue; Steve Monk at the University of Washington; Linda Davenport and Ron Narode at Portland State; Rochelle Kaplan at William Paterson College; Rebecca Corwin, Susan Jo Russell, and Andee Rubin at TERC (Technical Education Research Centers); Herb Ginsburg at Teachers College, Columbia University; Tom O'Brien at Southern Illinois State; Jan De Lange of the OWOC Project in the Netherlands; Steve Lerman at South Bank Polytechnic in London; and Paul Ernest at the University of Exeter, England, among others. Our ongoing conversations have often helped bring into relief issues in our work of which we had been insufficiently aware.

Finally, the community of teachers and administrators with whom we work has had a major role in shaping the program. At times, their direct suggestions (for example, that we offer semester-long mathematics courses) have moved the program in new directions. More importantly, our formal and informal discussions together provide grist for continuing reflection on our own goals and how to achieve them.

References

Ball, D. L. (1988). *The subject matter preparation of prospective mathematics teachers: Challenging the myths*. East Lansing: National Center for Research on Teacher Education, Michigan State University.

Ball, D. L. (1989). Research on teaching mathematics: Making subject matter knowledge part of the equation. In J. Brophey (Ed.), *Advances in research on teaching: Vol. 2. Teachers' subject matter knowledge and classroom instruction* (pp. 1–48). Greenwich, CT: JAI Press.

Ball, D. L. (in press-a). With an eye on the mathematical horizon: Dilemmas of teaching elementary school mathematics. *Elementary School Journal*.

Ball, D. L. (in press-b). Halves, pieces, and twoths: Constructing representational contexts in teaching fractions. In T. P. Carpenter, E. Fennema, & T. Romberg (Eds.), *Rational numbers: An integration of research*. Hillsdale, NJ: Lawrence Erlbaum Associates.

Ball, D. L., & Feiman-Nemser, S. (1988). Using textbooks and teachers' guides: A dilemma for beginning teachers and teacher educators. *Curriculum Inquiry, 18*, 401–423.

Belenky, M. F., Clinchy, B. N., Goldberger, N. R., & Tarule, J. M. (1986). *Women's ways of knowing*. New York: Basic Books.

Bennett, A., Maier, E., & Nelson, L. T. (1988). *Math and the mind's eye*. Salem, OR: Math Learning Center.

Brown, V. M. (1989). In the area of science, can constructivist teaching, based on children's own interests and investigations, provide a meaningful learning environment? *The Constructivist, 4*(4), 1–7.

Burns, M. (1987). *A collection of math lessons*. New Rochelle, NY: Cuisinaire Company of America.

Burns, M. (1991). *Math by all means: Multiplication grade 3*. New Rochelle, NY: Cuisinaire Company of America.

Burton, L. (1984). Mathematics thinking: The struggle for meaning. *Journal for Research in Mathematics Education, 15*(1), 35–49.

California State Department of Education. (1985). *Mathematics framework for California public schools, kindergarten through grade twelve*. Sacramento, CA: Author.

Carl, I. (1991, November). "Testing" can be an obstacle to change. *National Council of Teachers of Mathematics News Bulletin*, p. 3.

Carpenter, T. P., & Fennema, E. (in press). Cognitively guided instruction: Building on the knowledge of students and teachers. In W. Secada (Ed.), *Curriculum reform: The case of mathematics in the United States* [Special issue of *The International Journal of Educational Research*].

Chipman, S. F., Brush, L. R., & Wilson, D. M. (1985). *Women and mathematics: Balancing the equation.* Hillsdale, NJ: Lawrence Erlbaum Associates.

Clement, J., Narode, R., & Rosnick, P. (1981). Intuitive misconceptions in algebra as a source of math anxiety. *Focus on Learning Problems in Mathematics, 3*(3), 36–45.

Cobb, P. (1988). The tension between theories of learning and instruction in mathematics education. *Educational Psychologist, 23*(2), 97–103.

Cohen, D. K. (1988). Teaching practice: Plus que ça change. . . . In P. W. Jackson (Ed.), *Contributing to educational change: Perspectives on research and practice* (pp. 27–84). Berkeley, CA: McCutchan.

Cohen, D. K., & Ball, D. L. (1990). Policy and practice: An overview. In D. K. Cohen, P. L. Peterson, S. Wilson, D. Ball, R. Putnam, R. Prawat, R. Heaton, J. Remillard, & N. Wiemers, *Effects of state-level reform of elementary school mathematics curriculum on classroom practice* (Research Report 90-14; pp. 1–8). East Lansing: The National Center for Research on Teacher Education and The Center for the Learning and Teaching of Elementary Subjects, College of Education, Michigan State University.

Cohen, D. K., Peterson, P. L., Wilson, S., Ball, D., Putnam, R., Prawat, R., Heaton, R., Remillard, J., & Wiemers, N. (1990). *Effects of state-level reform of elementary school mathematics curriculum on classroom practice* (Research Report 90-14). East Lansing: The National Center for Research on Teacher Education and The Center for the Learning and Teaching of Elementary Subjects, College of Education, Michigan State University.

Confrey, J. (1985, April). *A constructivist view of mathematics instruction. Part I: A theoretical perspective.* Paper presented at the annual meeting of the American Educational Research Association.

Confrey, J. (1990). What constructivism implies for teaching. In R. B. Davis, C. A. Maher, & N. Noddings (Eds.), *Constructivist views on the teaching and learning of mathematics (Journal for Research in Mathematics Education* Monograph No. 4; pp. 107–124). Reston, VA: National Council of Teachers of Mathematics.

Corwin, R., & Russell, S. J. (1990). *Used numbers: Real data in the classroom.* Palo Alto, CA: Dale Seymour.

Crowley, M. (1987). The van Hiele model of the development of geometric thought. In M. M. Lindquist & A. P. Shulte (Eds.), *Learning and teaching geometry, K–12* (pp. 1–8). Reston, VA: National Council of Teachers of Mathematics.

Cuban, L. (1984). *How teachers taught: Constancy and change in American classrooms, 1890–1980.* New York: Longman.

Damarin, S. K. (1990). Teaching mathematics: A feminist perspective. In T. J. Cooney & C. R. Hirsch (Eds.), *Teaching and learning mathematics in the 1990s* (pp. 144–151). Reston, VA: National Council of Teachers of Mathematics.

Davis, P. J., & Hersh, R. (1980). *The mathematics experience.* Boston: Birkhauser.

Davis, R. B. (1966). Discovery in teaching mathematics. In L. S. Shulman & E. R. Keisler (Eds.), *Learning by discovery* (pp. 114–128). Chicago: Rand McNally.

Davis, R. B. (1984). *Learning mathematics: The cognitive science approach to mathematics education.* Norwood, NJ: Ablex.

Davis, R. B., Maher, C. A., & Noddings, N. (Eds.). (1990). *Constructivist views of the teaching and learning of mathematics.* Reston, VA: National Council of Teachers of Mathematics.

Driscoll, M. (1983). *Research within reach: Secondary school mathematics, a research-guided response to the concerns of educators.* Reston, VA: National Council of Teachers of Mathematics.

Duckworth, E. (1987). *"The having of wonderful ideas" and other essays on teaching and learning.* New York: Teachers College Press.

Ernest, P. (1991). *The philosophy of mathematics education: Studies in mathematics education.* London: Falmer.

Feiman-Nemser, S. (1983). Learning to teach. In L. S. Shulman & G. Sykes (Eds.), *Handbook of teaching and policy* (pp. 150–170). New York: Longman.

Fennema, E., & Leder, G. C. (Eds.). (1990). *Mathematics and gender.* New York: Teachers College Press.

Finkel, D. L., & Monk, G. S. (1983). Teachers and learning groups: Dissolution of the Atlas complex. In C. Bouton & R. Y. Garth (Eds.), *Learning in groups* (New Directions for Teaching and Learning No. 14; pp. 83–97). San Francisco: Jossey-Bass.

Fosnot, C. T. (1989). *Enquiring teachers, enquiring learners: A constructivist approach for teaching.* New York: Teachers College Press.

Fosnot, C. T. (1991). *Center for Constructivist Teaching teacher preparation project.* Paper presented to the Association of Teacher Educators, New Orleans, LA.

Foster, G., & Pellens, S. (1986). Teacher development as metamorphosis. *Educational Science Forum, 11,* 1–10.

Gilligan, C. (1982). *In a different voice.* Cambridge, MA: Harvard University Press.

Ginsburg, H. (1977). *Children's arithmetic: The learning process.* New York: Van Nostrand.

Ginsburg, H. (1986). *Children's arithmetic: How they learn it and how we teach it.* Austin, TX: Pro-Ed.

Ginsburg, H., & Kaplan, R. (1988). Overview—Math video workshop. New York: Teachers College, Columbia University.

Goldin, G. (1988). Affective representation and mathematical problem solving. In M. J. Behr, C. B. Lacampagne, & M. M. Wheeler (Eds.), *Proceedings of the Tenth Annual Meeting of the North American Chapter of the International Group for the Psychology of Mathematics Education* (pp. 1–7). DeKalb: Northern Illinois University.

Hall, G. E., Loucks, S. F., Rutherford, W. L., & Newlove, B. W. (1975). *Levels of use of the innovation: A framework for analyzing innovation adoption.* Austin, TX: The Research and Development Center for Teacher Education.

Hammerman, J., & Davidson, E. (in press). Homogenized is only better for milk. In G. Cuevas & M. Driscoll (Eds.), *Reaching all students in mathematics.* Reston, VA: National Council of Teachers of Mathematics.

Harel, G., Behr, M., Post, T., & Lesh, R. (1989). *Teachers' knowledge of multiplication and division concepts.* Unpublished manuscript.

Hart, L. (1991). Assessing teacher change. In R. G. Underhill (Ed.), *Proceedings of the Thirteenth Annual Meeting of the North American Chapter of the International*

Group for the Psychology of Mathematics Education (Vol. 2, pp. 78–84). Blacksburg: Division of Curriculum and Instruction, Virginia Polytechnic Institute and State University.

Hiebert, J. (Ed.). (1986). *Conceptual and procedural knowledge: The case of mathematics.* Hillsdale, NJ: Lawrence Erlbaum Associates.

Hutchinson, B. P., & Ammon, P. (1986). *The development of teachers' conceptions as reflected in their journals.* Paper presented at the annual meeting of AERA, San Francisco.

Joyce, B. (Ed.). (1990). *Changing school culture through staff development.* Alexandria, VA: Association for Supervision and Curriculum Development.

Joyce, B., & Showers, B. (1988). *Student achievement through staff development.* New York: Longman.

Kamii, C. (1985). *Young children reinvent arithmetic: Implications of Piaget's theory.* New York: Teachers College Press.

Kennedy, M. M. (1991). *An agenda for research on teacher learning.* East Lansing: Michigan State University, National Center for Research on Teacher Learning, Special Report.

Labinowicz, E. (1980). *The Piaget primer: Thinking, learning, teaching.* Menlo Park, CA: Addison-Wesley.

Lakatos, I. (1976). *Proofs and refutations.* Cambridge: Cambridge University Press.

Lampert, M. (1988a). The teacher's role in reinventing the meaning of mathematics knowing in the classroom. In M. J. Behr, C. B. Lacampagne, & M. M. Wheeler (Eds.), *Proceedings of the Tenth Annual Meeting of the North American Chapter of the International Group for the Psychology of Mathematics Education* (pp. 433–480). DeKalb: Northern Illinois University.

Lampert, M. (1988b). What can research on teacher education tell us about improving the quality of mathematics education? *Teaching and Teacher Education, 4*(2), 157–170.

Lampert, M. (1989, March). Arithmetic as problem solving. *Arithmetic Teacher,* pp. 34–36.

Leinhardt, G., & Smith, D. (1985). Expertise in mathematics instruction: Subject matter knowledge. *Journal of Educational Psychology, 77*(2), 247–271.

Lester, F. (1989, November). Mathematical problem solving in and out of school. *Arithmetic Teacher,* pp. 33–35.

Lester, J. (1987). Math journals: An individualized program. *The Constructivist, 2* (2), 1–7.

Lester, J. (1991). *Is the algorithm all there is?* Unpublished manuscript.

Maher, C. A. (1988). The teacher as designer, implementer, and evaluator of children's mathematics learning environments. *Journal of Mathematical Behavior, 6,* 295–303.

Maher, C. A., & Alston, A. (1990). Teacher development in mathematics in a constructivist framework. In R. B. Davis, C. A. Maher, & N. Noddings (Eds.), *Constructivist views on the teaching and learning of mathematics (Journal for Research in Mathematics Education* Monograph No. 4; pp. 147–166). Reston, VA: National Council of Teachers of Mathematics.

Mason, J., Burton, L., & Stacey, K. (1982). *Thinking mathematically.* London: Addison-Wesley.

Mathematical Association of America. (1991). *A call for change: Recommendations for the mathematical preparation of teachers.* Washington, DC: Author.

McDiarmid, G. W. (1990). *Tilting at webs of belief: Field experiences as a means of breaking with experience.* East Lansing, MI: The National Center for Research on Teacher Education.

McDiarmid, G. W., Ball, D. L., & Anderson, C. W. (1989). Why staying one chapter ahead doesn't really work: Subject specific pedagogy. In M. Reynolds (Ed.), *The knowledge base for beginning teachers* (pp. 193–205). New York: Pergamon Press.

Mokros, J. (1991). Meeting the challenge of mathematics assessment. *Hands On!, 14*(2), pp. 1, 18–19.

Mumme, J., & Weissglass, J. (in press). Improving mathematics education through school-based change. *Mathematics and Education Reform.*

Narode, R. (1988). *A constructivist program for college remedial mathematics: Design, implementation, and evaluation.* Unpublished doctoral dissertation, University of Massachusetts at Amherst.

National Council of Teachers of Mathematics. (1989). *Curriculum and evaluation standards for school mathematics.* Reston, VA: Author.

National Council of Teachers of Mathematics. (1991). *Professional standards for teaching mathematics.* Reston, VA: Author.

National Research Council. (1989). *Everybody counts: A report to the nation on the future of mathematics education.* Washington, DC: National Academy Press.

National Research Council. (1990). *Reshaping school mathematics: A framework for curriculum.* Washington, DC: National Academy Press.

Owens, D. T. (1988). Understanding decimal multiplication in grade six. In M. J. Behr, C. B. Lacampagne, & M. M. Wheeler (Eds.), *Proceedings of the Tenth Annual Meeting of the North American Chapter of the International Group for the Psychology of Mathematics Education* (pp. 107–113). DeKalb: Northern Illinois University.

Peterson, P., Fennema, E., & Carpenter, T. (1989, January/February). Using knowledge of how students think about mathematics. *Educational Leadership,* pp. 42–46.

Piaget, J. (1972). *Psychology and epistemology: Towards a theory of knowledge.* Harmondsworth, England: Penguin Books.

Piaget, J. (1977). *The principles of genetic epistemology.* London: Routledge and Kegan Paul.

Resnick, L. B. (1987). *Education and learning to think.* Washington, DC: National Academy Press.

Riddle, M. (1991). *Beyond stardom: Challenging competent math students in a mixed ability classroom.* Unpublished manuscript.

Ross, S. (1989, February). Parts, wholes, and place value: A developmental view. *Arithmetic Teacher,* pp. 47–51.

Russell, S. J. (in preparation). *Explorations in number, data, and space.* Palo Alto, CA: Dale Seymour.

Russell, S. J., & Corwin, R. B. (1991). Talking mathematics: "Going slow" and "letting go." In R. G. Underhill (Ed.), *Proceedings of the Thirteenth Annual Meeting of the*

North American Chapter of the International Group for the Psychology of Mathematics Education (Vol. 2, pp. 175–181). Blacksburg: Division of Curriculum and Instruction, Virginia Polytechnic Institute and State University.

Sarason, S. (1971). *The culture of the school and the problem of change.* Boston: Allyn and Bacon.

Sarason, S. (1990). *The predictable failure of educational reform: Can we change course before it's too late?* San Francisco: Jossey-Bass.

Schifter, D., & Simon, M. A. (1991). Towards a constructivist perspective: The impact of a mathematics teacher inservice program on students. In R. G. Underhill (Ed.), *Proceedings of the Thirteenth Annual Meeting of the North American Chapter of the International Group for the Psychology of Mathematics Education* (Vol. 1, pp. 43–49). Blacksburg: Division of Curriculum and Instruction, Virginia Polytechnic Institute and State University.

Schifter, D., & Simon, M. A. (in press). Assessing teachers' development of a constructivist view of mathematics learning. *Teaching and Teacher Education.*

Schoenfeld, A. H. (1989). Teaching mathematical thinking and problem solving. In L. B. Resnick & L. E. Klopfer (Eds.), *The 1989 Yearbook of the Association for Supervision and Curriculum Development* (pp. 83–103). Alexandria, VA: Association for Supervision and Curriculum Development.

Schram, P., Feiman-Nemser, S., & Ball, D. L. (1990). *Thinking about teaching subtraction with regrouping: A comparison of beginning and experienced teachers' responses to textbooks* (Research Report 89-5). East Lansing: The National Center for Research on Teacher Education, College of Education, Michigan State University.

Scott-Hodgetts, R., & Lerman, S. (1990). Psychological/philosophical aspects of mathematical activity: Does theory influence practice? In G. Booker, P. Cobb, & T. N. de Mendicuti (Eds.), *Proceedings of the Fourteenth PME Conference* (Vol. 1, pp. 199–206). Mexico City: Program Committee of the 14th Psychology of Mathematics Education Conference.

Shroyer, J., & Fitzgerald, W. (1986). *Mouse and elephant: Measuring growth.* Menlo Park, CA: Addison-Wesley.

Shulman, L. S. (1986a, February). Those who understand: Knowledge growth in teaching. *Educational Researcher,* pp. 4–14.

Shulman, L. S. (1986b). Paradigms and research programs in the study of teaching: A contemporary perspective. In M. C. Wittrock (Ed.), *Handbook of research on teaching* (3rd ed., pp. 3–36). New York: Macmillan.

Silver, E. A. (Ed.). (1985). *Teaching and learning mathematical problem solving: Multiple research perspectives.* Hillsdale, NJ: Lawrence Erlbaum Associates.

Simon, M. A. (1986, April). The teacher's role in increasing student understanding of mathematics. *Educational Leadership,* pp. 40–43.

Simon, M. A. (1989). *The impact of intensive classroom follow-up in a constructivist mathematics teacher education program.* Paper presented at the Annual Meeting of AERA, San Francisco. (ERIC ED313351)

Simon, M. A. (1990). Prospective elementary teachers' knowledge of division. In G. Booker, P. Cobb, & T. N. de Mendicuti (Eds.), *Proceedings of the Fourteenth PME*

Conference (Vol. 3, pp. 313–320). Mexico City: Program Committee of the 14th Psychology of Mathematics Education Conference.

Simon, M. A. (1991). The Construction of Elementary Mathematics (CEM) Project. National Science Foundation grant No. TPE–9050032.

Simon, M. A., & Schifter, D. E. (1991). Towards a constructivist perspective: An intervention study of mathematics teacher development. *Educational Studies in Mathematics, 22*(5), 309–331.

Steffe, L. P., Cobb, P., & von Glasersfeld, E. (1988). *Construction of arithmetical meanings and strategies.* New York: Springer-Verlag.

Thompson, A. (1984). The relationship of teachers' conceptions of mathematics and mathematics teaching to instructional practice. *Educational Studies in Mathematics, 15*(2), 105–127.

Tymoczko, T. (1986). *New directions in the philosophy of mathematics.* Boston: Birkhäuser.

von Glasersfeld, E. (1983). Learning as a constructive activity. In J. C. Bergeron & N. Herscovics (Eds.), *Proceedings of the Fifth Annual Meeting of the North American Chapter of the International Group for the Psychology of Mathematics Education* (pp. 41–69). Montreal: Université de Montréal, Faculté de Science de l'Education.

von Glasersfeld, E. (1990). An exposition on constructivism: Why some like it radical. In R. B. Davis, C. A. Maher, & N. Noddings (Eds.), *Constructivist views on the teaching and learning of mathematics (Journal for Research in Mathematics Education* Monograph No. 4; pp. 19–30). Reston, VA: National Council of Teachers of Mathematics.

Weinberg, D., & Parker, M. B. (1988). Teacher as learner, learner as teacher. In M. J. Behr, C. B. Lacampagne, & M. M. Wheeler (Eds.), *Proceedings of the Tenth Annual Meeting of the North American Chapter of the International Group for the Psychology of Mathematics Education* (pp. 319–326). DeKalb: Northern Illinois University.

Weissglass, J. (1991). Teachers have feelings: What can we do about it? *Journal of Staff Development, 12*(1), 28–33.

Wood, T., Cobb, P., & Yackel, E. (1991). Change in teaching mathematics: A case study. *American Educational Research Journal, 28*(3), 587–616.

Yaffee, L. (1991). *Pictures at an exhibition: A math phobic confronts fear, loathing, cosmic dread and thirty years of math education.* Unpublished manuscript.

Index

About the Authors

DEBORAH SCHIFTER has worked with the SummerMath Programs at Mount Holyoke College since their inception in 1982, becoming the director of SummerMath for Teachers in 1988. She has also worked as an applied mathematician and has taught elementary, secondary, and college level mathematics. She has a BA in liberal arts from Saint John's College, Annapolis, an MA in applied mathematics from the University of Maryland, and an MS and a PhD in psychology from the University of Massachusetts. In March 1993 she will move to the Education Development Center, Inc., in Newton, MA, where she will be working at the Center for the Development of Teaching.

CATHERINE TWOMEY FOSNOT is Associate Professor of Education and director of the Center for Constructivist Teaching, Teacher Preparation Project at Southern Connecticut State University in New Haven. In 1986, while on leave from SCSU, she worked on the SummerMath for Teachers project at Mount Holyoke college. She is past president of the Association for Constructivist Teaching, the author of *Enquiring Teachers, Enquiring Learners* (also published by Teachers College Press), and in 1984 she received the ERIC/ECTJ Young Scholar Award for her writing on the topic of constructivism and educational technology.